AIRCRAFT ENGINE MAINTENANCE AND SERVICE

AIRCRAFT ENGINE MAINTENANCE AND SERVICE

by COLONEL ROLLEN H. DRAKE, B.S., M.A.

RESEARCH ENGINEER, PILOT AND GROUND SCHOOL INSTRUCTOR
FORMERLY: INSTRUCTOR, VOCATIONAL AND AVIATION SUBJECTS, LOS
ANGELES CITY SCHOOLS; CHIEF, AIR AGENCY UNIT AND CIVILIAN PILOT
TRAINING SPECIALIST, CIVIL AERONAUTICS ADMINISTRATION; SUPPLY
SPECIALIST, OFFICE OF THE QUARTERMASTER GENERAL

Originally published in 1950

ISBN: 978-1-940001-40-1

The Aviation Collection
by
Sportsman's Vintage Press
2015

This book is dedicated to

MY BELOVED WIFE

without whose inspiration and help

this book could not have been written

PREFACE

THE aircraft engine, like the human body or any other complicated mechanism made up of moving parts and designed to produce power, is subject to many little ills. These troubles develop through normal wear under operating conditions. While the body can repair its own minor damages or injuries, an engine is not able to "get well" by itself. Neglected troubles in the aircraft engine will eventually lead to more serious trouble or failure. Any failure of the engine in flight is serious and if the failure is complete the airplane must be landed immediately.

It is the duty of those in charge of maintenance and service of the engine to keep it in condition that it will always develop full power under operating conditions. It is obvious that the maintenance and service of an aircraft engine requires the highest type of skilled personnel. All members of maintenance and service crews—from the highest skilled mechanic down to the one who washes the airplane wings—are responsible for the proper maintenance of the engine to assure safe flight operation. The mechanic in charge of the aircraft engine is perhaps required to be expert in more lines than any other type of mechanic. He must be machinist, welder, electrician, instrument man, and an inspector of the highest type.

This book is written to bring before laymen, the students, teachers and certificated aircraft engine mechanics the fundamentals of aircraft maintenance and service. Each branch of the field of maintenance and service is adequately covered. This book is not designed to be used as a text in the fundamentals, theory and operation of the aircraft engine but it does include much of this material. It covers completely the various types of inspections which are required to keep engines in safe operating condition.

This book includes top overhaul, disassembly, overhaul inspection, repair and replacement of engine parts, assembly and engine run-in after overhaul. The text is profusely illustrated with photographs and drawings which have a direct relationship to the subject matter. This book furnishes all material necessary in the way of theory to qualify one for the Civil Aeronautics Administration's aircraft engine mechanic certificate. It also meets the requirements for a text for formal class work in aircraft engine maintenance and service. It will also be of great value to members of maintenance crews and to certificated mechanics. While it has been designed definitely as a reference book for mechanics in the field and for classroom use, it can be understood readily by the casual reader. It will also serve for use in College Vocational Courses, Trade Schools, Junior Colleges, High Schools, Aviation Ground Schools and for Rehabilitation groups.

Perhaps the most outstanding feature of this book is the non-technical, simple language in which it is written; the author has tried to avoid the use of formulas, graphs, confusing tables, obscure footnotes and any other material which cannot be understood clearly. Particular attention has been given to inspection and approved repairs as required by the Civil Aeronautics Administration.

The author wishes to express his grateful appreciation to the following, who have so kindly furnished material which has been of assistance in the preparation of this book: the U. S. Office of Education; Civil Aeronautics Administration; the Department of Education of the various states, particularly New York, Pennsylvania, Utah and Virginia; Engineer and Research Corporation; Ranger Aircraft Engines; Jacobs Aircraft Engine Company; Wright Aeronautical Corporation; Piper Aircraft Corporation; Lycoming Division, The Aviation Corporation; Allison Division, General Motors Corporation; General Electric Company; Gladen Products Division, Los Angeles Turf Club, Inc.; Kinner Motors, Inc.; Pesco Products Company; Eaton Manufacturing Company; Taylorcraft Aviation Corporation; A C Spark Plug Division, General Motors Corporation; Eisemann Corporation; Delco-Remy Division, General Motors Corporation; Eclipse-Pioneer Division, Bendix Aviation Corporation; Stinson Division, Consolidated-Vultee Aircraft Corporation; Minneapolis Honeywell Regulator Company; Glenn L. Martin Company; Douglas Aircraft Company; The Torrington Company; S K F Industries, Inc.; Norma-Hoffman Bearing Corporation; Timken Roller Bearing Corporation; Candler-Hill Corporation; Marvel-Schebler Division, Borg-Warner Corporation; Scintilla Magneto Division, Bendix Aviation Corporation: The Electric Auto-Lite Cor-

PREFACE

poration; Continental Aircraft Engines; Hamilton Standard Propellers; Curtis Wright Corporation; Aero Products Division, General Motors Corporation; and Thomson Industries, Inc. The author particularly wishes to thank those of the above who have read and edited parts of the manuscript.

The author wishes also to thank his many friends who have so generously contributed their advice and assistance. He also wishes particularly to express his gratitude to Earle R. Hough for his excellent work in the preparation of many of the drawings in this book and to Mildred Pickrel and Alma Franklin for their untiring patience and cooperation in the preparation of the manuscript.

The frontispiece shows a three-quarter front view of a 7-cylinder, radial, aircraft engine. The cylinder baffles have been installed. It is reproduced by courtesy of Jacobs Aircraft Engine Company.

R. H. D.

TABLE OF CONTENTS

INTRODUCTION

Every shop or factory or place where persons work has in force certain safety rules and safety practices to prevent injuries to those who are employed. Much education is devoted to health and to the prevention of the injuries and sicknesses to which the human body is subject. Good health measures require that small cuts or injuries receive proper care to prevent their developing into serious ailments.

The modern aircraft engine is as complicated, perhaps, as the human body. Many engines have more separate parts than does the body, and each part is subject to its own minor ills and wears out even under normal use. A part of the human body, such as a hand or foot, may be out of service and yet the body goes on and performs most of its functions. Any part of the engine which fails may cause the engine to stop or may lead to the failure of parts closely connected with that part which fails. It is not enough to determine that a part has failed, but it is absolutely necessary that, if the engine is to carry the airplane and its cargo safely through the air, indications of failure must be detected and remedies applied to prevent failure.

Any serious injury to the human body requires the services of a physician or a skilled surgeon. The aircraft engine, however, requires the services of the skilled aircraft-engine mechanic to correct even its minor ailments.

A ground vehicle may be driven with comparative safety although many parts are at the point of failure. For instance, an automobile may be driven at slow speeds with tires that are in danger of going flat. Almost any part of the car may be ready to give way, and even a part of the main structure may fail when the car is driven at slow speeds without serious danger to the occupants of the car or the rest of the vehicle itself. If a car breaks down, the engine stops, or a part of the body gives way, it is only necessary to push or pull the vehicle to the side of the road until it can be repaired or towed away.

If an airplane engine fails, it is necessary that the plane be landed at once regardless of the surface of the ground underneath. The pilot selects the best available landing place but the one sure thing is that he must land, and land at once.

The pilot of an airplane would never consider taking off unless he

Fig. 1. The engine cowling, power plant and the nose wheel of a light airplane. (Courtesy Engineering and Research Corporation)

was sure that the airplane engine was functioning properly and, insofar as humanly possible, as a result of proper inspections, sure that no part showed the least sign of failure.

Most people do not take as good care of their bodies as one must of an airplane engine. They do not carry out all the rules of safety and health, and many are apt to neglect minor injuries. The little illnesses of the engine must be taken care of at once. The engine, unlike the human body, cannot get well by itself. Every little failure will become worse with continued use and will eventually lead to the failure of the engine at a time perhaps when it is needed the most.

There is no distinct line between service, maintenance, overhaul, and repair. Service may consist of such simple operations as removing excess

Fig. 2. The power plant installation of a light airplane. (Courtesy Engineering and Research Corporation)

oil or dirt from the engine, draining the carburetor trap, tightening nuts, or renewing safety wire.

However, the removal of dirt and grease from the engine, which is a service, may assist in prolonging the life of the engine and thus fall into the field of maintenance.

Maintenance may be thought of as keeping the engine in such condition that it is always as good as when it was first built. Even with the most careful maintenance, the engine cannot be maintained in a completely new condition, as there is always normal wear and deterioration of the various parts. The points of the spark plugs gradually burn away, and every bearing, no matter how carefully lubricated, eventually wears out. It is, however, the duty of the maintenance crew to maintain the engine in such condition that it will always develop its full horsepower and pull the airplane through the air to its destination.

The pilot and passengers must always be assured that, no matter what else may happen, the engine will not fail. Maintenance may include such operations as renewing the finish which protects the engine

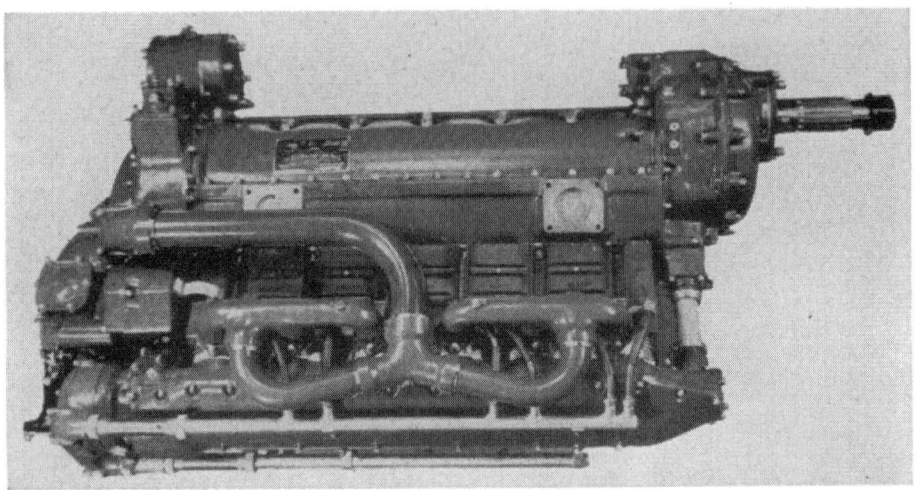

Fig. 3. The right side view of an inverted "V" type of in-line, radial engine. (Courtesy Ranger Aircraft Engines)

against corrosion, tightening nuts, or placing cotter pins or other small parts which show signs of wear or failure. Minor repairs may be classified as maintenance as well as any other more or less simple operation which is necessary to keep the engine in safe flight condition.

Overhaul and repair may, in a broad sense, be considered maintenance. These operations are designed to keep the engine, as nearly as possible, in a new condition, or at least in a condition which allows 100 per cent normal operation. Overhaul does not necessarily include repairs. Overhaul may simply include the disassembly of the engine to determine whether or not all of the parts are in such shape that it is safe to operate the aircraft engine under flight conditions. Overhaul to determine the condition of the engine, requires inspection by a person thoroughly trained in inspection procedures. Reassembly requires the services of a skilled mechanic, for the expert and conscientious assembling of an aircraft engine is in many ways as delicate as the assembling of a fine watch.

Repairs to an aircraft engine, in the real sense of the word, are not common. Few parts of an airplane engine may be repaired, if seriously damaged. Damaged parts are usually replaced by new parts furnished by the manufacturer. Many parts of the aircraft structure may be repaired by splicing, patching, or the replacing of a few rivets. Minor repairs to the aircraft structure may include the patching of a damaged fabric, the repair of a hole in the metal or plywood skin covering the assembly or fuselage, or may consist of such major repairs as splicing a main spar or welding in a new part of the fuselage.

Actual repairs to engine parts usually consist of such operations as the removal of burrs or scratches from the teeth of gears, smoothing out shallow scratches in the cylinder barrel, or smoothing up a rough bearing. There is no such operation in engine repair as welding a broken crankshaft or patching a cracked pistonhead.

The duties of the aircraft-engine mechanic are largely those of maintenance and service except at overhaul periods.

Service consists of the complete inspection of all parts of the engine which may be visually inspected and the maintaining of all parts in proper adjustment insofar as possible.

Maintenance and service must necessarily include such minor replacements and adjustments as are necessary to keep the engine in safe flight condition, and do not include major overhauls. Maintenance and service cannot be clearly defined as to whether the operation will be performed by a certificated aircraft-engine mechanic or by another person. Many operations which fall into the field of maintenance and service may be performed by persons other than a certificated mechanic if performed under the supervision of such a mechanic. Other items of service and maintenance may be performed by a properly trained ground crew.

It is the responsibility, however, of the mechanic in charge to see that each item of maintenance and service is properly performed at the proper time. Most items of repair and overhaul must be performed by a certificated mechanic.

The experts of the Civil Aeronautics Administration have set up complete rules, regulations, and techniques covering the inspection, maintenance, repair, and service of aircraft engines in order that they may be maintained in safe flight condition.

Before flight, the Civil Aeronautics Administration requires that the engine be thoroughly checked by a qualified person. This preflight check includes all inspections which may be made visually and covers all parts of the power plant, including the propeller. These inspections also include the condition of the cowling to see that it is in place and properly secured.

Certain inspections must be made at regular intervals by certificated engine mechanics. It is the responsibility of every pilot, mechanic, and

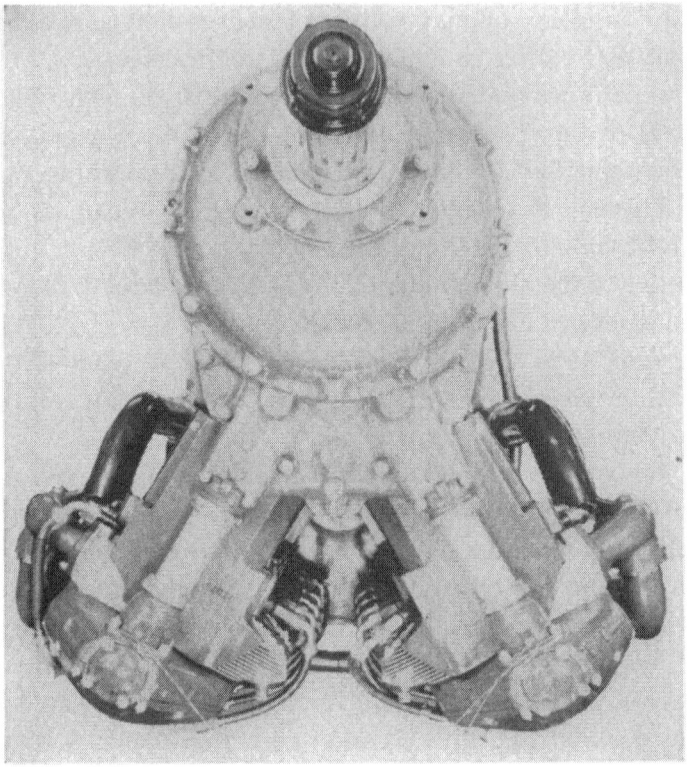

Fig. 4. A front view of a 12-cylinder, inverted, "V" type of in-line aircraft engine. (Courtesy Ranger Aircraft Engines)

owner to be sure that the engine meets all of the requirements for safe operation before flight.

Much depends upon the knowledge, skill, and conscientiousness of the person charged with maintenance, inspection, and service. It is not only important to know how to perform the operations necessary, but it is just as important to be able to tell when they should be performed and when they have been performed properly.

Many operations pertaining to maintenance, repair, and overhaul have been explained in detail in this text, but it is assumed that the fundamental techniques of maintenance, repair, service, and inspection are understood by the mechanic. Most operations of this kind require the services of a certificated aircraft-engine mechanic, and this text should assist such a person in the performance of his duties.

It is impossible for any one text of this kind to cover every type of aircraft engine completely. An attempt has been made in this text to give the fundamentals of maintenance, service, overhaul, and repair as required by the Civil Aeronautics Administration for the engines used in civil aircraft.

The operations described in this text, insofar as possible, have been chosen as those which are typical of any aircraft engine.

This book is not intended to be a fundamental text in the theory, operation, and construction of aircraft engines, but to cover as thoroughly as possible the field of maintenance and service.

FUEL, LUBRICATION, IGNITION, AND COOLING AND INDUCTION SYSTEMS

This chapter has to do with general servicing procedures of 1. Fuel Systems 2. Lubrication Systems 3. Ignition Systems 4. Cooling and Induction Systems.

The procedures given in this chapter are, more or less, general and the mechanic or airman in charge of the inspection or service should refer constantly to the "Manufacturers' Maintenance and Service Manual" for the particular aircraft upon which he is working.

1. FUEL SYSTEMS

Fuel systems are designed to supply a sufficient amount of fuel to the engine and carry the reserve fuel necessary for extended flights. The fuel system is designed to clean the fuel, removing any solid impurities. A satisfactory fuel system must be designed to operate under all conditions of altitude and various flight positions of the aircraft.

The person in charge of service should pay particular attention to the fuel gauges to see that they are in proper operating condition and indicate at all times the exact amount of fuel in the tanks.

Each time the aircraft is serviced, all fuel lines, cut-off cocks, connections and flexible joints should be carefully examined for leakage. All fuel lines should be examined for signs which might indicate failure.

Refueling is, perhaps, the service operation most frequently performed in aircraft service. As soon as an aircraft returns from a flight, the tank should be checked to determine the amount of fuel remaining. The entire fuel system should be given a visual check to determine whether leaks might have developed during the flight. The tank should be immediately filled. This prevents condensation of moisture from the air in the tank.

FUEL AND INDUCTION SYSTEMS

The proper octane rating should always be checked for each aircraft. On some of the larger airplanes, fuel of a higher octane rating may be used for take-off and climb. This fuel, of course, is placed in a separate tank in the system. These airplanes may use a fuel of somewhat lower octane rating for cruising.

If the aircraft is equipped with a number of tanks and all of the tanks are not to be filled, the proper tanks should be filled to maintain balance. On light airplanes, particularly, dirt or foreign matter is apt to get into the fuel tank of the airplane through the filler cap opening while fueling. Dirt and dust should be carefully cleaned from around the filler cap, and the members of the ground crew should be sure that dirt does not fall from their clothing, gloves, or any part of the refueling apparatus into the fuel tank.

The passage of fuel through the hose and nozzle may generate static electricity. The airplane may itself have an accumulated charge of electricity due to its passage through the air or from the ignition system itself. When the fuel nozzle is placed in the filler opening to the tank, a spark of electricity may jump, causing an explosion. To prevent this, a ground wire, preferably grounded to an underground fuel tank, is fastened by means of a spring clamp to a metal part of the aircraft to allow the discharge of any static electricity. In some light aircraft, due to modern rubber mountings on the engine, the engine and the fuel system may be quite well insulated from the rest of the aircraft. The ground safety wire should then be grounded to some such part as the propeller hub nut. If a funnel and chamois strainer are used to strain the gasoline, the funnel should be grounded. The hose discharge nozzle should be attached to a ground wire which is also grounded. The spark may jump just as the discharge nozzle is removed from the filler vent. Of course, no open lights of any kind and no smoking should be allowed in the vicinity of an airplane while it is being fueled. Only explosion proof flashlights should be used around aircraft. Proper fire extinguishers should be at hand whenever fueling operations are taking place. If it is possible to avoid it, an airplane should never be refueled in a hangar or other enclosure. This operation should be performed in the open air. Members of the service crew should avoid breathing gasoline fumes which are quite poisonous. Gasoline in contact with the skin may cause a burn as serious as fire or steam. Clothing that has been wet in gasoline should be removed as soon as possible and the skin washed with water and soap.

The Carburetor. The modern carburetor usually needs but little attention between engine overhauls. However, at regular engine inspection periods, the carburetor and fuel lines should be inspected and checked. Any signs of leakage, or even of slight leakage which may be indicated by staining with fuel dyes, should be corrected. All nuts, bolts, studs, and safetyings should be inspected visually after each flight or at least once each day. Air scoops and manifold connections should be checked regularly. All throttle and carburetor controls should be inspected periodically for ease of operation, slack or play, and for proper safetying. Carburetor sediment traps should be drained daily before the first flight. All screens and strainers should be cleaned at regular intervals. The float chamber should be drained regularly. Before draining the float chamber and carburetor, the feed pipe should be shut off and a check should be made to determine whether or not any water drains from the carburetor. After draining the carburetor, the feed valve should be opened enough to allow fuel to flow freely from the feed pipes. All drain plugs and connections should be checked for tightness and proper safetying. When a carburetor is not functioning properly, the trouble may often be found by a process of elimination.

A mechanic must keep in mind that the carburetor must function properly under a number of different operating conditions, such as idling, acceleration, cruising, and full power. A carburetor should be checked while the engine is idling and, to see that the engine responds properly, when the throttle is opened. The engine should accelerate smoothly and rapidly to full power. It is sometimes necessary to test the carburetor under flight conditions to determine its proper functioning.

Air-Fuel Mixture. Proper mixture is important. A mixture that is too lean may cause backfiring, detonation, increased engine temperature, irregular firing, and an excessively hot manifold. A mixture that is too rich may cause a cool running engine, irregular running, and smoke in the exhaust gases. The smoke caused by a too-rich mixture is usually blue or black in color and should not be confused with the white smoke caused by excessive oil consumption. A flooded engine may sometimes be indicated by black smoke from the exhaust.

The carburetor is often blamed for trouble caused by faults outside the carburetor itself. A lean mixture or a rich mixture may prevent the engine from developing its full-rated horsepower. Air leaks in the intake manifold have the same effect as a lean mixture adjustment in the carburetor. Poor ignition or sticking valves may result in the same symptoms

10

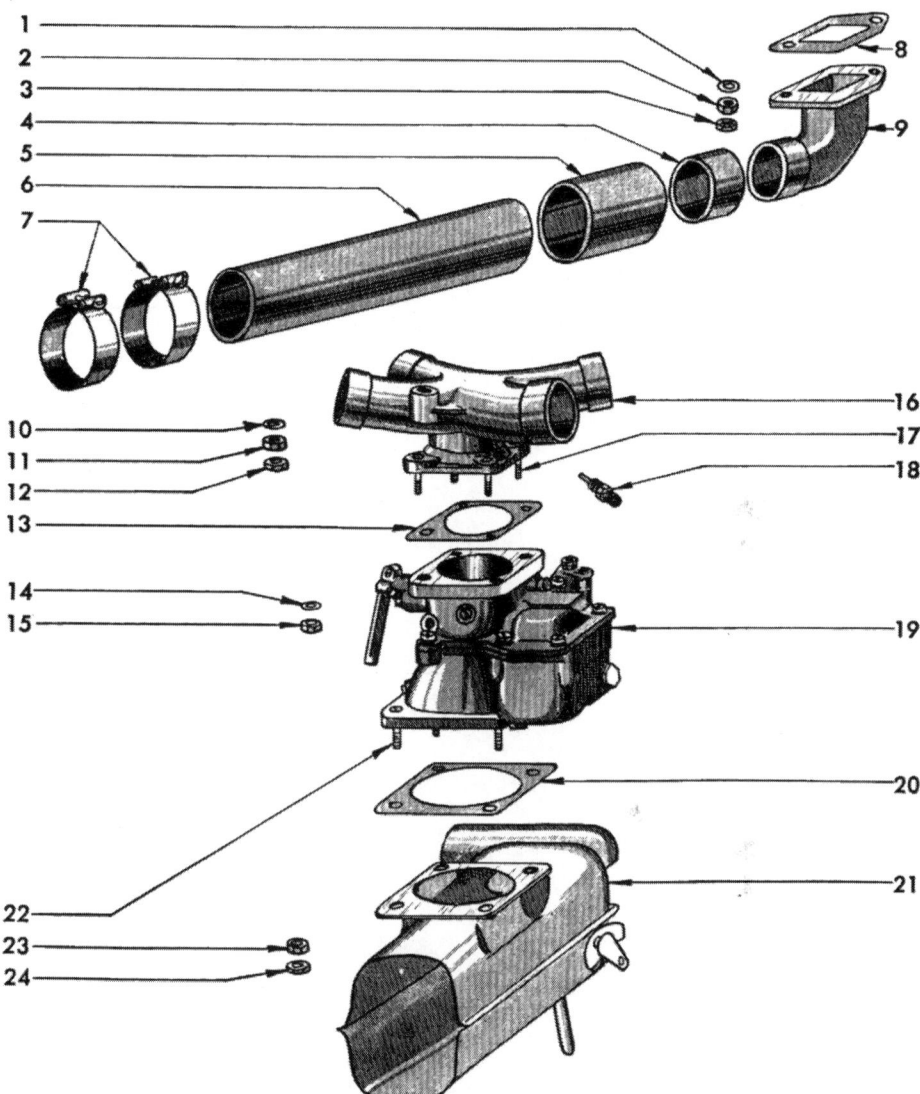

Fig. 5. The Carburetor and air-intake system for a light, 4-cylinder aircraft engine showing: (1) Washer, (2) nut, (3) palnut, (4) hose, (5) hose, (6) intake pipe, (7) intake-pipe hose clamp, (8) intake elbow gasket, (9) intake elbow, (10) intake manifold to crankcase washer, (11) intake manifold to crankcase nut, (12) intake manifold to crankcase palnut, (13) carburetor to intake manifold gasket, (14) carburetor to intake manifold washer, (15) carburetor to intake manifold nut castle; (16) intake manifold assembly, (17) carburetor to intake manifold stud, (18) primer jet, (19) carburetor, (20) gasket, (21) carburetor air intake, (22) stud, (23) nut, (24) lock nut. (Courtesy Continental Aircraft Engines)

11

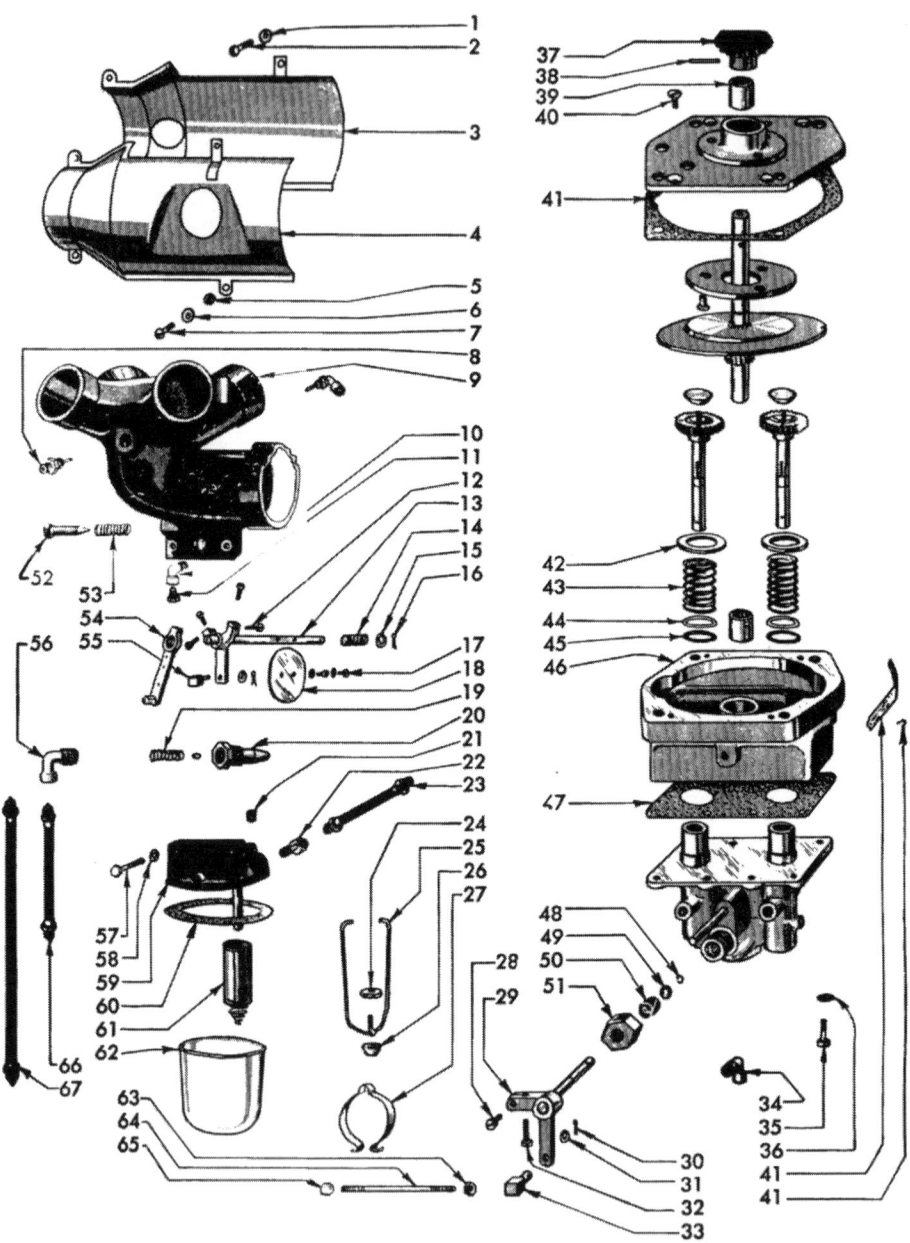

Fig. 6. An exploded view of fuel injector equipment showing (1) lockwasher, (2) screw, (3) air-intake housing assembly, left half, (4) air-intake housing assembly, right half, (5) nut, (6) lockwasher, (7) screw, (8) primer jet, (9) intake manifold, (10) elbow, (11) nut and sleeve assembly, (12) screw, (13) air-control, valve-shaft assembly (fuel injector), (14) air-control, valve-shaft spring, (15) air-control valve-shaft spring washer, (16) cotter pin, (17) screw, (18) throttle valve, (19) plunger spring, (20) engine intake elbow jet assembly, (21) plug, (22) coupling, female, (23) sediment bowl to fuel injector tube assembly, (24) bowl screw

12

as those produced by a lean mixture. Backfiring and the igniting of the mixture in the intake manifolds may be caused by preignition or faulty timing. Weak compression may cause overheating with the same symptoms as a lean mixture. The adjustment of the carburetor from rich to

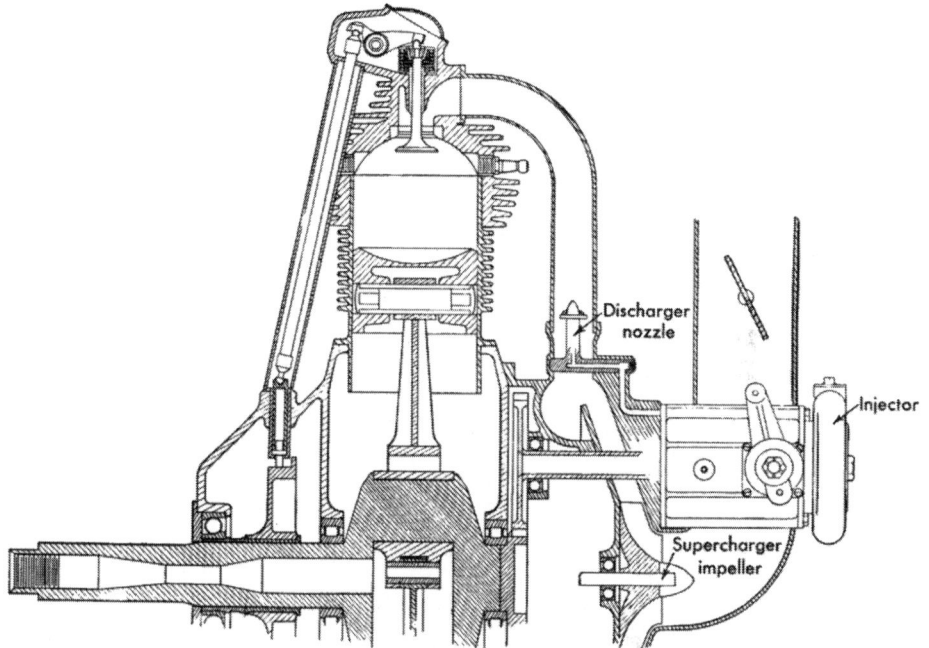

Fig. 7. A schematic drawing showing a fuel-injector system, side view.

lean and back to rich again will indicate whether or not the trouble is within the carburetor itself. The above faults will not be corrected by mixture changes. While alternately leaning and enriching the mixture, the engine speed should be checked to determine whether the speed varies with changes in the mixture. If the engine does not respond to

nut, (25) bowl clamp and screw assembly, (26) sediment bowl seat, (27) clip, (28) screw, (29) metering control valve-shaft assembly, (30) cotter pin, (31) washer, (32) screw, (33) injector lever to throttle-lever-rod end, (34) fuel injector vent elbow, (35) screw, (36) lock washer, (37) fuel-injector gear, (38) pin, (39) bushing, (40) screw, (41) gasket, (42) plunger-spring seat, (43) plunger spring, (44) plunger-spring washer, (45) gasket, (46) fuel-injector unit complete, (47) gasket, (48) circlip, (49) washer, (50) packing, (51) gland nut, (52) injector air bleed screw, (53) injector air bleed screw spring, (54) lever assembly-throttle, (55) injector lever to throttle-lever-rod end, (56) injector jet elbow, (57) screw, (58) washer, (59) sediment bowl cover, (60) sediment bowl gasket, (61) strainer assembly, (62) glass sediment bowl, (63) nut, (64) injector lever to throttle-lever rod, (65) nut, (66) fuel-injector vent tube assembly, (67) fuel injector to intake elbow tube assembly. (Courtesy of Continental Aircraft Engines)

changes in the mixture, the trouble is probably in some other place than the carburetor.

Faulty rings allowing oil to pass the pistons will give much the same indications as an overrich mixture. The engine should be run at fairly high speed (1100 to 1250 r.p.m.) for a short time until the excess oil

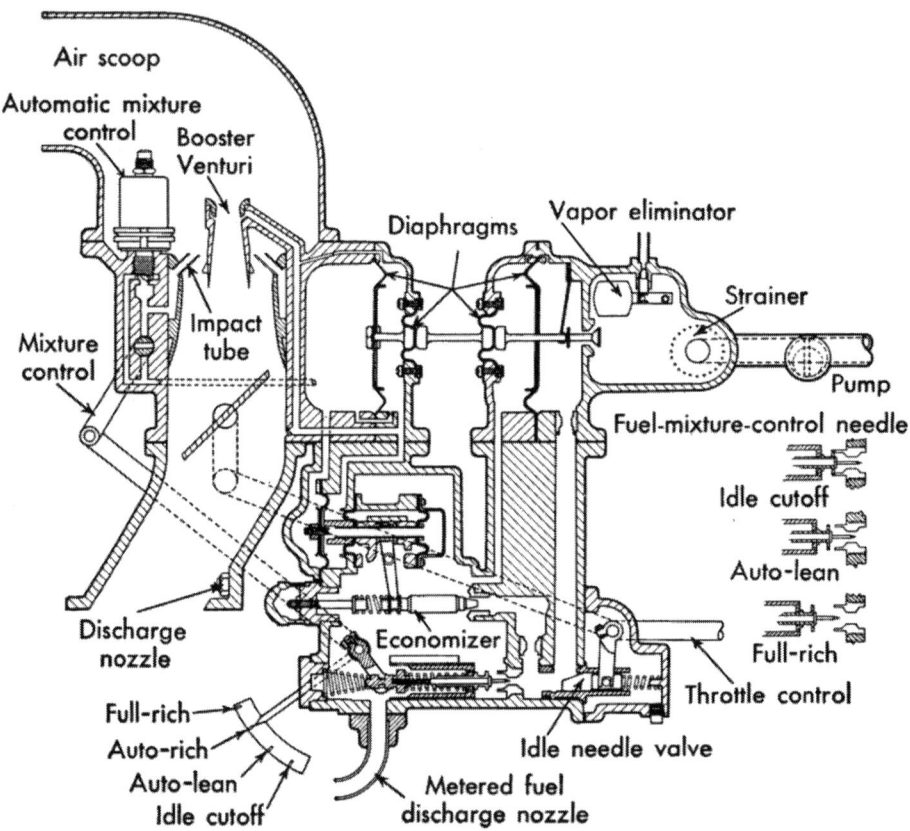

Fig. 8. A pressure-type carburetor.

and fuel have been cleared from the engine and then the throttle slowly closed. If the engine immediately begins to miss, after being slowed down, a condition of both rich mixture and "oil pumping" may be indicated. If the engine runs for some time after it has been slowed down before beginning to miss, oil pumping is indicated.

The presence of water in the carburetor may make the engine difficult to start and may cause the engine to stop when speeded up. Water is heavier than gas and will settle to the bottom of the carburetor and may not be picked up at low engine speeds. At high speeds, water being drawn into the air-fuel mixture from the carburetor will cause rough running and backfiring, a symptom which is similar to that produced by a lean mixture.

It is impossible to describe all the symptoms which occur with every make of carburetor, but it is important that the mechanic recognize

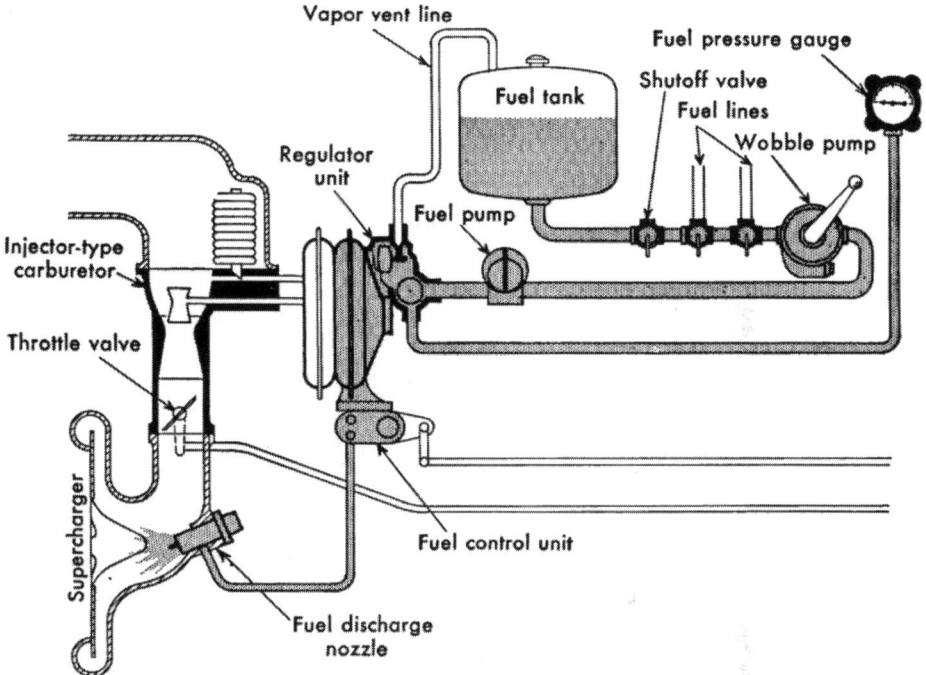

Fig. 9. A schematic drawing of an injection carburetor system.

the more common ones. The mechanic should follow carefully the operating and service instructions furnished by the manufacturer for each particular make of carburetor.

Rich Mixture. A rich mixture may be caused by (1) the fuel pressure being too high; (2) the float level being adjusted too high; (3) the mixture control being set to an overrich position; (4) oversize or loose jets; (5) air bleeds, clogged or partially clogged; (6) obstructions in the air scoop and air intake; (7) an economizer setting which causes the opening of the economizer too early; (8) a leak in the accelerator pump or valve; (9) a leak in the primer-pump valve allowing fuel to be sucked into the induction system through the priming system. In general, a rich mixture is indicated by blue or black smoke from the exhaust. After dark, the flame from the exhaust may be dark red in color.

15

Uneven running or galloping of the engine with reduced power and excessive fuel consumption indicates an overrich mixture. An overrich mixture exerts a cooling effect on the engine. The engine will not usually overheat until the mixture is so rich that the engine power is considerably reduced.

Lean Mixture. Lean mixtures may be caused by low fuel pressure or the mixture control being set in the lean position. The fuel level may be too low and the jets too small or partially clogged, or the accelerator pump and economizer may not be functioning properly. Air leaks or cracks in the induction system between the carburetor and engine, clogged fuel lines or strainers, clogged vents in the fuel tanks, or a partially closed fuel-control valve will probably cause a lean mixture. If the engine is not properly warmed up, it may choke and stop when the throttle is rapidly opened. A partial vapor lock may cause a lean mixture. A cool induction system or excessively cold air entering the carburetor may also produce a lean mixture. After dark, a lean mixture produces long, light blue and yellow flame from the exhaust stack. Loss of power, overheating, detonation, preignition, backfiring, and spitting are all probable indications of a lean mixture which may cause any or all of them.

A correct mixture should produce a short, dark blue flame from the exhaust stacks after dark. Of course, the same types of flames are produced in daylight but are very difficult to see.

High Fuel Consumption. Excessive fuel consumption may be caused by (1) operating the engine at full throttle; (2) the jets being too large or loose in the carburetor body; (3) the air bleeds being too small; (4) the fuel pressure being too high; (5) a leaking or soaked carburetor float which allows the fuel level to become too high; (6) over.speeding the engine when a fixed-pitch propeller is used; (7) improper adjustment of the carburetor; or (8) a leaking primer system.

Vapor Locks. Vapor locks may be caused by (1) vertical, short, or upward bends in the fuel line; (2) fuel lines that are placed too close to hot spots; (3) any arrangement of the fuel lines which allows pockets in which vapor may be trapped; or (4) fuel containing an excess of highly volatile fractions. Fuel locks or vapor locks are indicated by a sudden stoppage of the engine, particularly in hot weather, or failure of the engine to start in hot weather after having been stopped a short time. This condition is particularly apt to cause stoppage of the engine during take-off.

16

Air Leaks. Air leaks in the induction system may be the result of cracks in the induction housing, leaking gaskets, or loose joints in the intake-pipe arrangement. Leaking gaskets in the carburetor are usually indicated by the engine idling fast or not at all. Air leaks may sometimes be heard as a high whistling sound while the engine is idling.

Each engine is designed to operate with a specified carburetor or carburetors. The engine manufacturer's recommendations as to the proper carburetor should be followed. When installing a carburetor on an aircraft engine, it is the responsibility of the mechanic to see that the proper carburetor is used. The specification will be stamped on the carburetor name plate. Jet sizes and all specifications are given for each carburetor. It is important that all carburetor and throttle controls be carefully installed and safetied. The engine manufacturer usually specifies the correct idling speed for the engine which is usually about 550 r.p.m.

Idling Speed. The idling speed is one which will not produce excessive vibration. Most carburetors have an idling-speed adjustment. The idling mixture control should be so adjusted as to get the best power at idling speed. Care should be taken that the idling mixture is not so rich that the engine will "load up" when idling over long periods of time. When carburetors are equipped with a mixture control, the idling mixture should be adjusted until all cylinders are firing evenly. All cylinders should fire evenly when the throttle is opened normally to an engine speed of about 1000 to 1100 r.p.m. Uneven firing during acceleration indicates that the mixture is too lean and that there is not a large enough acceleration charge of fuel. After adjusting the idling mixture, the throttle should be closed normally to determine that there is no tendency for the engine to stop when the r.p.m. are reduced to idling speed.

When carburetors are equipped with mixture controls, the proper setting may be determined by the following process. Allow the engine to idle with the mixture adjustment at its middle position until conditions have become stabilized. Move the main mixture-control lever to the lean position and, if the engine speeds up, it is an indication that the idling mixture is too rich. The idling adjustment should then be moved to a leaner position and the engine again checked by moving the main mixture-control lever to the lean position. If the speed definitely decreases, it is an indication that the idling mixture is too lean. The idling mixture setting is correct when there is a tendency for the speed to drop off slightly when the main mixture control is moved from the full-rich to the full-lean position. If the engine is equipped with a manifold-pressure gauge, the correct idling mixture is the one which produces the lowest manifold-pressure reading.

The following are operating and maintenance instructions for a light aircraft-engine carburetor.

The carburetor is made up of two major units: a cast aluminum throttle body and bowl cover, and a cast aluminum fuel bowl and air entrance.

Idle System. With the throttle fly slightly open to permit idling, the suction or vacuum above the throttle on the manifold side is high. Little air passes through the Venturi at this time and hence, with low suction on the main nozzle, it does not discharge fuel. The high suction is above the throttle valve. Fuel from the fuel bowl passes through the fuel-channel power jet and into the main nozzle bore. It then passes through the idle supply opening in the main nozzle through the idle fuel orifice in the idle tube. Here it is mixed with air which has entered the idle tube through the primary and secondary idle air vents. This rich mixture of fuel and air passes upward through the idle tube where it is finally drawn into the throttle barrel through the primary

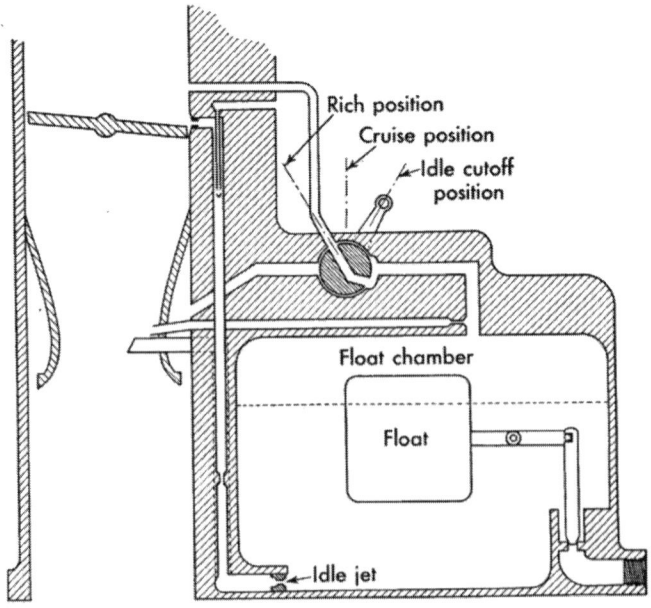

Fig. 10. A typical idler system and idle cut-off used on modern carburetors.

idle delivery opening. The flow may be regulated by the idle adjusting needle or the idle adjustment lever. The small amount of air passing the throttle valve mixes with it, forming the air-fuel mixture for idling the engine. The idle adjusting needle controls the ratio of the air and fuel in the mixture supplied to the throttle barrel, and therefore controls the quality of the idle mixture. Turning the needle counterclockwise, away from its seat, makes the mixture richer and turning the needle clockwise, towards its seat, leans the idle mixture.

On idle, some air is drawn past the throttle barrel below the throttle valve through the secondary idle delivery opening and blends with the idling mixture being delivered to the engine.

As the throttle is opened, the secondary idle delivery begins to add to the idling mixture. This fuel adds to the idle delivery to prevent the engine from stalling before the main nozzle starts to operate.

Metering. All fuel delivery at idling speed, and also at steady propeller speeds up to approximately 1000 r.p.m., is from the idle system. At approximately 1000 r.p.m., the main nozzle begins to function. Due to decreased suction on the idling system, this system stops adding fuel

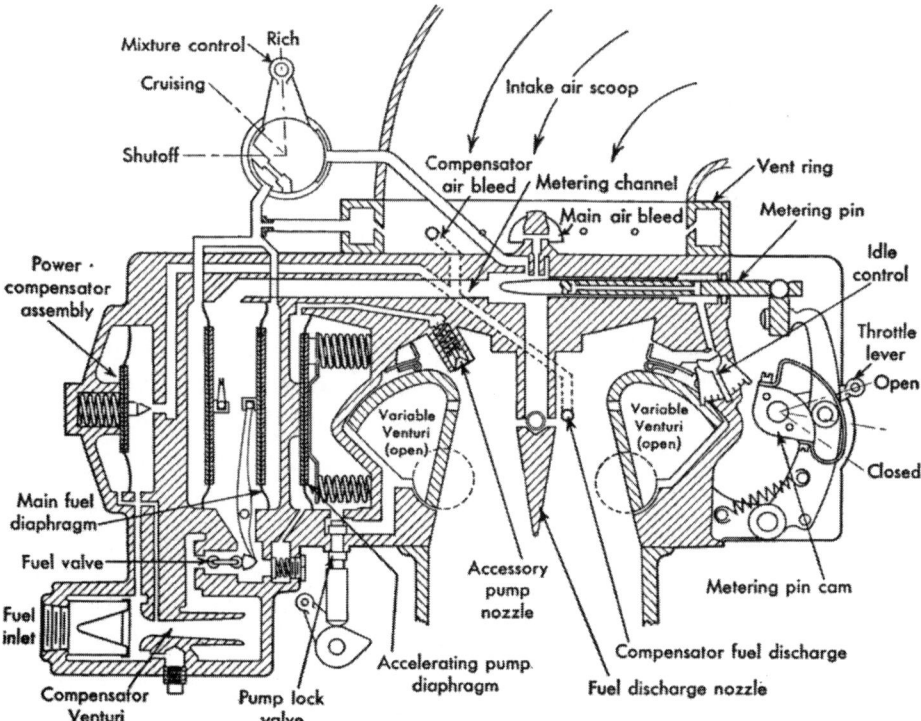

Fig. 11. A variable Venturi carburetor showing "full throttle" position.

to the mixture at about 1400 r.p.m. Most of the fuel delivery from that point on to the highest speed is from the main nozzle. The idle system and the main nozzle are connected with each other by the idle supply opening. The amount of fuel delivered from either the idle system or main nozzle depends on whether the suction is greater on the idle sys-

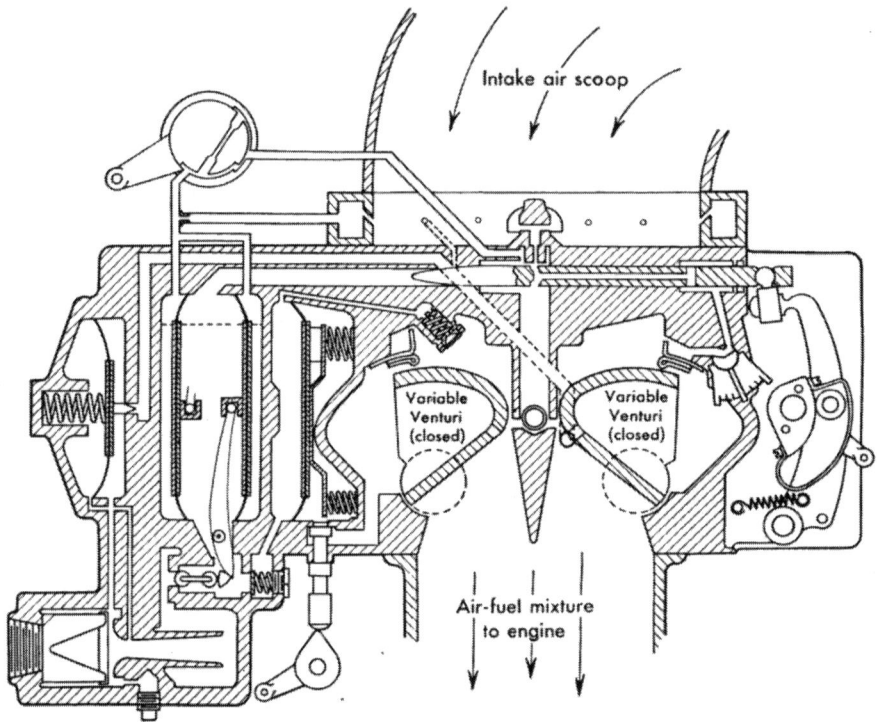

Fig. 12. A variable Venturi carburetor showing "idling" position.

tem or main nozzle. This suction is governed by the throttle-valve position and engine load. The main nozzle feeds at any time the throttle is open sufficiently to place the engine under a great enough load to decrease the manifold pressure. Under such conditions of low manifold pressure at the throttle valve, the main nozzle feeds instead of the idle system. This is brought about by the suction on the main nozzle due to the action of the Venturi.

Main Nozzle. The main nozzle is supplied with fuel which passes from the fuel bowl through the metering sleeve and the fuel channel. The fuel then passes upward through the nozzle bore where it is mixed with air drawn from the nozzle air vent and nozzle bleed holes. It is then discharged from the nozzle outlet into the mixing chamber as an air and fuel emulsion. Air passing through the nozzle air vent sweeps fuel from the nozzle well and nozzle bore under very low suction and therefore satisfies any sudden demand for nozzle fuel delivery. The nozzle and air-vent entrance are provided with an air-vent screen to prevent clogging of vents with bugs and foreign matter.

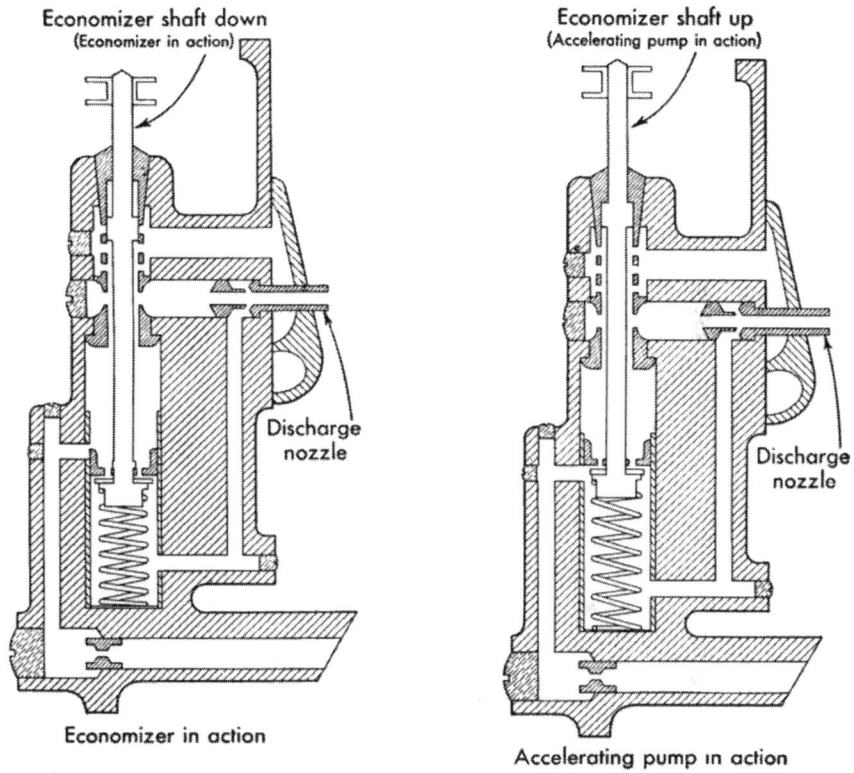

Fig. 13. An economizer and accelerating pump.

Back-Suction Economizer. In order to provide a leaner mixture to obtain full economy in the cruising range, the back-suction economizer is provided. This device has no moving parts and consists of connecting channels from the throttle barrel to the sealed fuel bowl. The economizer hole opens into the throttle barrel and negative pressure acts upon the fuel in the fuel bowl. This back suction diminishes the fuel flow to the nozzle and idle system.

Accelerating Pump. The accelerating pump discharges fuel only when the throttle valve is moved towards the open position. It provides additional fuel to keep in step with the sudden inrush of air into the manifold when the throttle is opened. By means of an accelerating-pump lever connected to the throttle shaft, the accelerating-pump plunger is moved downward when the throttle is opened. This forces fuel into the mixing chamber of the carburetor.

Upon closing the throttle, the accelerating-pump plunger moves upward. This refills the accelerating-pump chamber by drawing fuel

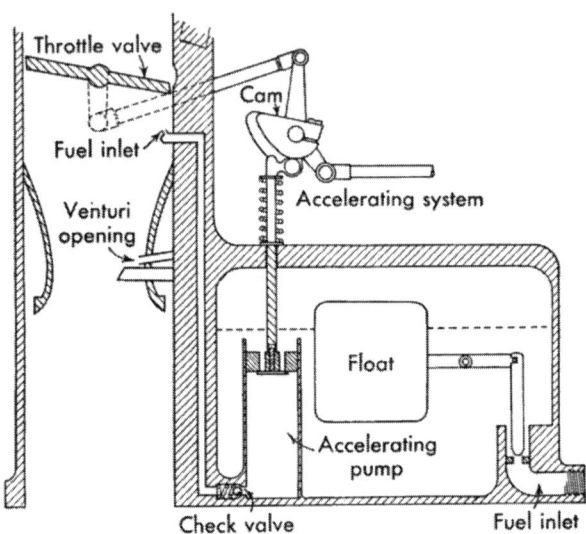

Fig. 14. A typical accelerating carburetor pump.

from the fuel bowl through the pump inlet screen and pump inlet check valve. On any quick opening of the throttle, the pump follow-up spring yields and thus prolongs the pump discharge sufficiently to prevent choking the engine with fuel.

As a precaution to prevent fuel from being drawn into the mixing chamber when the accelerating pump is inoperative, as it is at any constant throttle position, a carburetor pump discharge check-valve assembly is provided. This check valve is usually of the spring-loaded ball type. The accelerating-pump lever usually has an adjustment by which the amount of accelerating fuel may be regulated.

Adjustment of Carburetor. With a newly installed engine or when tuning up an engine, it is often found necessary to readjust the carburetor. The following is a recommended procedure for adjusting the carburetor. With the engine thoroughly warmed up, set the throttle stop screw so that the engine idles at approximately 550 r.p.m. Turn the idle adjusting needle out slowly until the engine "rolls" from richness, then turn the needle in slowly until the engine "lags" or runs irregularly from the lean mixture. This step will give an idea of the idle adjustment range and of how the engine operates under these extreme idle mixtures. From the lean setting, turn the needle out slowly to the richest mixture that will not cause the engine to roll or run unevenly. This adjustment will, in most cases, give a slower idle speed than a slightly

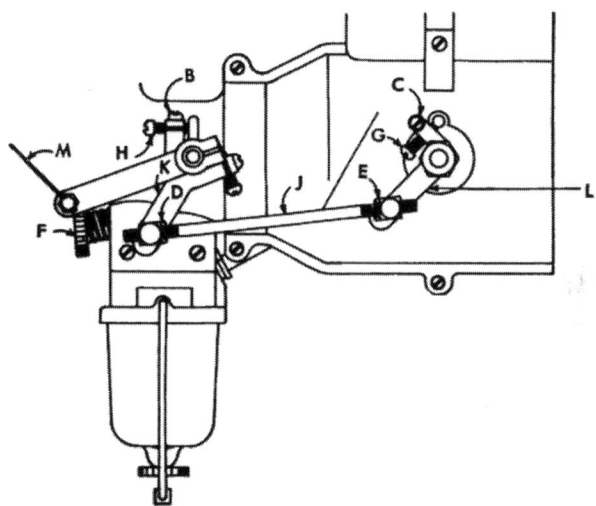

Fig. 15. A diagram of the fuel and air adjustment for Excello and High injectors in which B = Idle speed adjusting screw lock; C = Idle fuel adjusting screw lock; D = Link rod lock nut; E = Link rod lock nut; F = Idle air adjusting screw; G = Idle fuel adjusting screw; H = Idle speed adjusting screw; J = Link rod; K = Throttle control lever; L = Fuel control lever; M = Control wire to cockpit. (Courtesy Continental Aircraft Engines)

leaner adjustment with the same throttle stop screw setting, but will give the smoothest idle operation. A change in idle mixture will change the idle speed, and it may be necessary to readjust the idle speed with the throttle stop screw to the desired point. The idle adjusting needle should be from ¾ to 1 turn from its seat to give a satisfactory idle mixture.

CAUTION: Care should be taken not to damage the idle needle seat by turning the idle adjusting needle too tightly against the seat, as damage to this seat will make it difficult to obtain a satisfactory idle adjustment.

Float Height. The float height is set at the factory and can be checked by removing the throttle body, bowl cover, and float assembly, and turning it upside down. The power setting of the float should be made by measuring the distance from the bowl-cover gasket to the closest surface of the float in accordance with the manufacturer's recommendation. If the carburetor is provided with two floats, be sure to check both, making sure that the floats are parallel to the bowl-cover gasket.

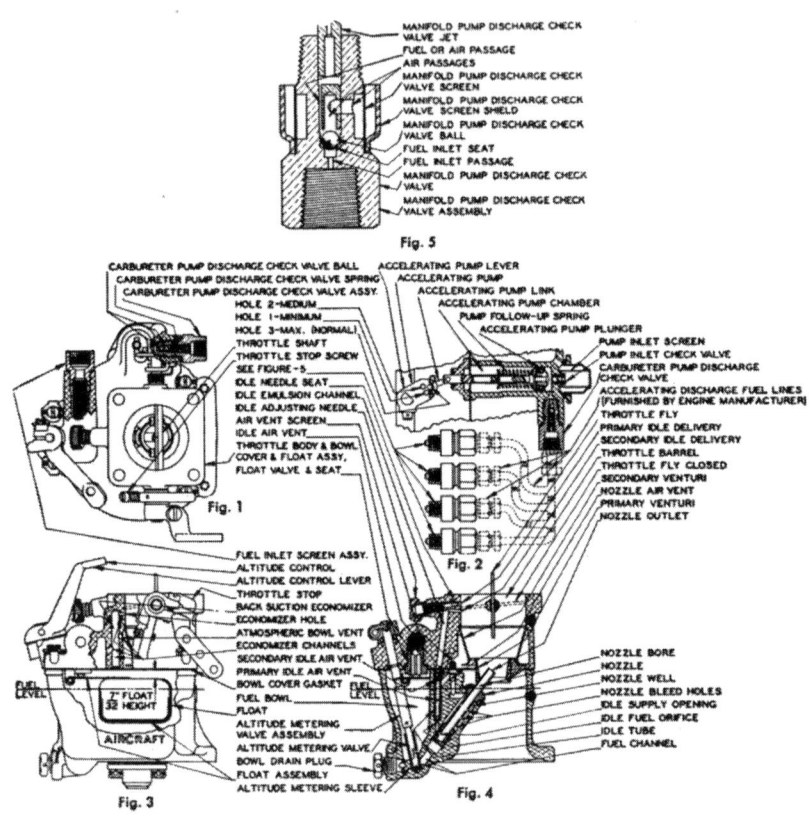

Fig. 16. Diagrams showing the construction and parts of an aircraft carburetor. (Courtesy Marvel Schebler Division, Borg-Warner Corporation)

Starting a Cold Engine. The throttle should be opened and closed two or three times depending on the air temperature. If the air temperature is low, more "pumping" of the throttle is required. The throttle should be "cracked" at approximately $\frac{3}{32}$ in. from the throttle stop screw. With the throttle in this position, the engine should be turned over two or three times before the ignition is turned on. This will draw a finely emulsified mixture of air and fuel through the manifold into the combustion chamber. The ignition should then be turned on, and the engine should start on the next turnover. With the throttle stop

³⁄₃₂ in. from the throttle stop screw, there should be sufficient throttle opening to keep the engine running. The carburetor is calibrated to give the richest mixture at this throttle opening and, therefore, a cold engine will run the smoothest with the throttle in this position. For

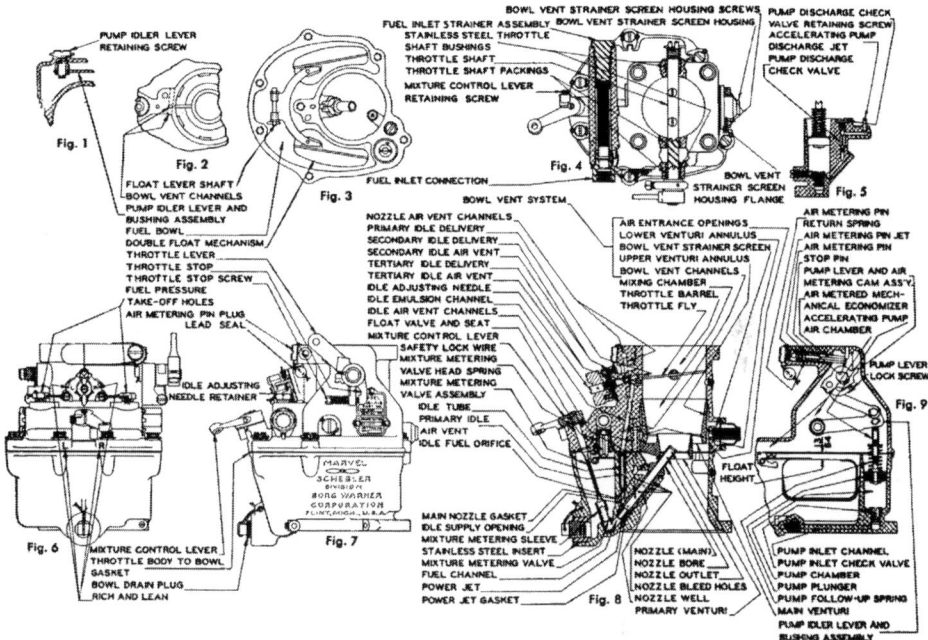

Fig. 17. Diagrams showing the parts of an aircraft carburetor. (Courtesy Marvel Schebler Division, Borg-Warner Corporation)

this reason, the engine should be allowed to warm up for several minutes before opening the throttle farther.

Starting a Hot Engine. To start a warm or hot engine, pull the throttle stop against the throttle stop screw in the idling position. If the engine has just been shut off, turn on the ignition, and the engine should start on the first turn but, if the engine has been shut off for several minutes, it may be necessary to turn the engine over once or twice before turning on the ignition. A warm or hot engine should start and continue to run with the throttle in the idling position.

2. LUBRICATION SYSTEMS

The lubricating system of the engine is as important as any other part of the aircraft because if it fails, a forced landing becomes necessary.

There are two main classifications of lubricating systems: the wet sump and the dry sump.

In the wet sump system the main oil supply is carried in the crankcase or in a special sump attached to the engine. This system is used chiefly on aircraft engines of low horsepower.

The dry sump system became necessary when engines were inverted and the stationary radial type of engine came into use.

In engines of high horsepower, it is necessary that some provision be made to cool the oil. Oil is an important factor in engine cooling. The oil radiator should always be closely examined during service periods.

In the dry sump engine the oil supply is carried in a tank entirely separate from the engine. The oil is furnished to the engine by means of a pressure pump. The oil after having passed through the bearings in the engine is removed from the engine by a scavenger pump and returned through the oil radiator to the oil supply tank.

Each airplane engine has its own particular system, the servicing of which must be carried out in accordance with the "Manufacturer's Manual."

Servicing the Lubrication System. If it were possible to have perfectly lubricated engines, wear would be reduced to practically zero. Lubrication of machinery, particularly of rapidly moving parts such as are contained in an aircraft engine, is often not very effective. The damage being done to the machine while in operation is often not apparent to the operator. A poorly lubricated bearing may be subjected to excessive wear without the operator's being aware that the lubrication of that part has partially failed.

In maintenance and service, lubrication is one of the most important items and must be carefully considered by the service crew as well as the operator of the aircraft. The grade of oil recommended by the

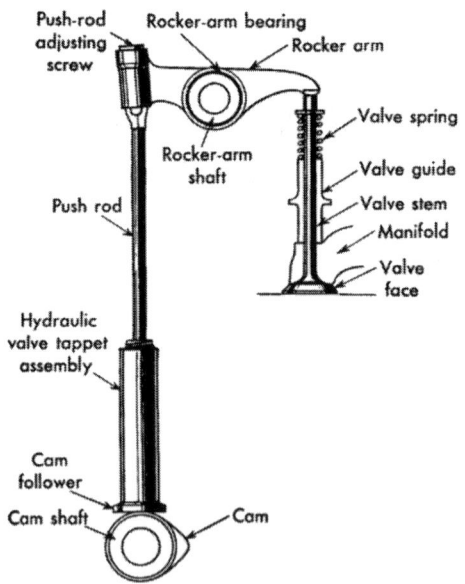

Push-rod adjusting screw

Rocker-arm bearing

Rocker arm

Rocker-arm shaft

Push rod

Valve spring

Valve guide

Valve stem

Manifold

Valve face

Hydraulic valve tappet assembly

Cam follower

Cam shaft

Cam

Fig. 18. A drawing to show the various parts of a valve train. (Courtesy Eaton Manufacturing Company)

manufacturer should always be used. The oil level should be checked every time the aircraft is refueled and, if necessary, oil should be added to the supply to maintain it at the proper level. The external oil lines should be visually inspected for any indication of leaks each time the aircraft is refueled.

The proper functioning of the oil-pressure gauges and oil systems should be checked each time the engine is started, as well as during its

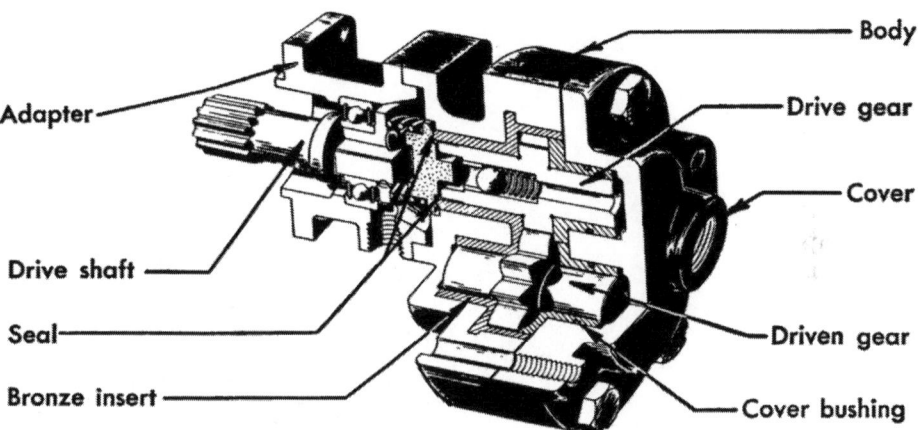

Fig. 19. A gear-type oil pump. (Courtesy Pesco Products Company)

operating time. Oil should be drained at the intervals recommended by the manufacturer or operating agency. Every time the oil is changed, the engine sump, oil cooler, oil-tank sump, and external lines should be drained and thoroughly flushed out. Any standard flushing fluid may be used.

If the engine is equipped with a magnetic plug, the plug should be carefully examined for metallic particles clinging to the magnet. All screens in the lubrication system should be removed and carefully cleaned. It is well to remove and thoroughly clean the by-pass relief valve. Particles of foreign matter which might become lodged in the relief valve may interfere with its proper functioning. If the lubrication system is equipped with a Cuno filter, it should be inspected and cleaned at regular intervals. Every time the airplane is refueled, two or three turns should be given the Cuno filter handle. It is well to remove the oil pump itself for a thorough cleaning at the time the oil is changed. All oil screens and plugs should be replaced and carefully safetied before filling the supply tank with oil after draining.

At regular intervals, which may be at the regular engine overhaul periods, the oil tank and oil cooler should be removed and thoroughly cleaned and inspected. The tank should be inspected, not only for leaks, but for any accumulation of foreign matter. Soft aluminum tanks may be repaired by welding with an oxy-acetylene welding torch. If the tanks are made from strong aluminum alloys, the repair should be made by replacing an entire panel or by patching. Patches should be riveted on, and a proper sealing compound should be applied to prevent leaking. Special cleaning compounds should be used to remove sludge and carbon deposits from the tank, fuel lines, or oil coolers.

Fig. 20. A scavenge oil pump showing the magnetic strainer, oil seal, and splined shaft for the fuel pump. (Courtesy Ranger Aircraft Engines)

When tanks are to be welded, all traces of oil and oil fumes should be removed from the tank. The tank should not be welded while in place in the aircraft. To remove all traces of oil which might cause an explosion during the welding process, the tank should be thoroughly cleaned with kerosene and then with gasoline. Live steam should then be used to remove the last traces of gasoline, and the tank washed with carbon tetrachloride.

The above process may be used to clean the tanks thoroughly before reinstalling them at the regular overhaul period. Before reinstalling, the tank should be tested with 3 or 4 lb. of air pressure to discover any leaks. While the tank is under pressure, soapy water should be applied with a brush to the entire external surface. Any small leaks will be shown in the form of bubbles. The tank should be carefully rinsed of the soap and water before reinstalling. If tanks are to be placed in storage, they should be thoroughly cleaned and lubricated inside and out with a coat of light oil.

Leaks in oil coolers may be repaired with a high grade tin-lead solder. To remove the tubes, two soldering irons having ends which just fit into the ends of the tube should be used. The irons are inserted in the two ends of the tube at the same time and, as the solder melts, the tube may be pushed out. New tubes may then be inserted and soldered into place.

Oil coolers and oil tanks should not be cleaned with soap or soap compounds nor with any solution containing lye. Lye affects many metals, and caustic compounds will combine with oil forming soap which may cause foaming in the lubrication system.

No attempt should be made to adjust spring-loaded relief valves, thermostatic valves, or viscosity valves as they are not adjustable. If they are not operating properly, they should be replaced. The proper functioning of these valves should be tested by special test equipment. Visual inspection, unless the part is actually broken, gives no indication of whether or not it is operating properly.

Lubrication-System Troubles. When the engine is first started, the oil-pressure gauge should be immediately observed to see whether or not it is operating. The oil-pressure gauge should indicate full pressure immediately.

Fig. 21. A side view of a scavenge oil pump showing oil strainer and magnetic plug. (Courtesy Ranger Aircraft Engines)

If the oil pressure does not show on the gauge within 30 sec., the engine should be stopped and not started again until the trouble has been located. After the engine has been warmed up, there should be no fluctuation of the indicator needle.

To determine the cause for a lack of oil pressure, low oil pressure, or fluctuating pressure, the mechanic should check the following: (1) the oil supply, (2) the oil pump, (3) the oil suction line, (4) the oil-relief valve, (5) the oil screens. There may be no oil in the supply tank. The oil pump itself may not be operating, which may be due to lack of prime. Air may have become trapped between the pump and the relief valve mechanism, or the oil suction line may be filled with air. The oil-relief valve should be removed and examined for proper adjustment. The oil suction line should be disconnected, allowing the oil to drain freely from the line, and the pump itself should be filled with oil. The engine should be turned over by hand until the oil begins to show pressure on the pressure gauge. The oil screens may be clogged. These should be removed, inspected, and cleaned. The oil suction lines should be examined for leaks. Low oil pressure may be due to improper adjustment of the relief valve or failure of the relief valve spring.

Low Oil Pressure.   Worn bearings may be allowing the oil to escape too freely from the bearings. Low oil pressure from this cause usually becomes more apparent as the engine warms up and the oil becomes thinner. If low oil pressure occurs while the engine is hot, the oil cooler should be checked and the oil-pressure-relief valve adjusted. If low pressure continues to be indicated, worn bearings or a leak in the system are probably the cause.

Irregular Oil Pressure.   Fluctuations of oil pressure may be caused by foaming in the oil supply tank. The scavenged oil from the engine normally contains a large number of air bubbles. This oil should be returned to the tank over a suitably installed baffle to prevent the excessive mixing with oil already in the tank. Clogged vent lines may cause foaming, as will water in the oil.

If foaming occurs, the oil should be drained from the system and the system thoroughly cleaned before being filled with fresh oil.

High Oil Pressure.   Variations in oil pressure may be caused by changes in speed and oil temperatures. Excessively high pressures may be shown when the engine is first started. This pressure should not be cause for alarm unless the pressure remains high after the engine has reached its normal operating temperature. Excessively high oil pressure may be due to the failure of the pressure-relief valve.

There is a normal drop in oil pressure as the oil temperature increases. In order to be sure that the oil-pressure gauge is indicating the correct oil pressure, the gauge line should be filled with light oil from time to time. As the gauge line becomes filled with heavy oil, an incorrect pressure may be indicated.

High Oil Temperature.   Excessively high oil temperature may be caused by a low oil supply or a failure of the oil-cooling system. A low grade of oil may allow the engine to become overheated, thus overheating the oil. The oil-temperature gauge indicates the engine temperature indirectly by measuring the temperature of the oil being drained from the engine. Any sudden rise in engine temperatures under operating conditions should be checked for its cause immediately. The improper lubrication of any part of the engine, particularly the bearings, will cause high oil temperatures. Dirty and contaminated oil, worn piston rings, and piston rings that have been incorrectly installed may cause high oil temperatures.

To check the cause for excessive oil temperatures, the vents of the oil system, strainers, and coolers should be examined for proper functioning. The oil lines should be examined to be sure they are open and that parts of the oil lines have not become restricted due to denting, kinking, or any other cause.

High temperatures may be caused by a failure of the scavenger pump, allowing too much oil to accumulate in the crankcase. One cause for failure of the scavenger pump is lack of prime. The scavenger pump should be examined to determine whether it is properly primed. This may be done by disconnecting the engine oil-outlet line and feeding oil to the scavenger pump through the oil-outlet connection. While the oil is being fed to the pump, the

Fig. 22. A breather assembly which is attached to the main crankcase. (Courtesy Ranger Aircraft Engines)

engine should be turned backward by hand until the pump is primed. If the scavenger pump is operating properly, the trouble is probably due to clogged oil lines or strainers, although excessively high temperatures may thin the oil to such an extent that the scavenger pump does not operate efficiently.

Metal particles of any kind found in the lubrication systems are an indication of failure somewhere within the engine itself. The engine should be immediately disassembled for overhaul. Before reinstalling an engine under these conditions, the entire lubrication system must be completely disassembled and all parts thoroughly cleaned to remove any metal particles which may have become lodged in any of its parts. The oil tanks should be cleaned thoroughly with particular attention given to small cracks or crevices along the sump baffles and joints where small metal particles might lodge. All oil lines should be thoroughly washed and then swabbed out with clean rags. After the swabbing, kerosene or a similar substance should be poured through the lines to remove any lint. All drain valves and fittings should be cleaned in the same manner as the oil lines. The thermostatic viscosity valves and Cuno filters should be disassembled sufficiently to allow thorough cleaning. It is usually recommended because of the difficulty of removing metal particles from the oil cooler that oil coolers should be discarded when metal particles have been found in the lubrication system.

3. THE IGNITION SYSTEM

The ignition system should not be disassembled as long as it is functioning properly. However, at regular intervals, the ignition system should be visually inspected to be sure that all connections are tight and properly fastened. The magneto ground wire terminals should be checked for tightness daily or before flight.

At regular intervals the breaker points should be examined. Any oil or grease present in the breaker mechanism should be removed with a small brush or a rag dipped in unleaded gasoline. The points should be checked to see that they are properly adjusted and that no pitting or

Fig. 23. A cutaway view of a magneto mounted on an engine. (Courtesy of Ranger Aircraft Engines)

burning has taken place. A faulty condenser may cause excessive burning of the contact points. The breaker points should not be filed while in place, but should be removed and supported in a suitable block while being filed or honed with a stone. The points should be timed to open when the piston in the No. 1 cylinder is at its full advance firing position.

The magneto does not usually need lubricating except at regular overhaul periods. Usually the proper functioning of the magnetos is not obtained until the engine is running at approximately 70 per cent of its normal rated r.p.m. The failure of one magneto may cause detonation when the engine is operating near its maximum power. Running on one magneto should cause a dropping off of from 50 to 150 r.p.m.

Spark-plug failure is one of the common causes of ignition trouble. In case of ignition trouble, the spark plugs should be examined unless the trouble is obviously located in some other part of the system. The plugs should be tested in a regular spark-plug tester, and the points properly adjusted according to the manufacturer's recommendations. Irregular running may be caused by the failure of one or two spark plugs. The spark plugs that are not firing may be located by shorting out each plug with a screw driver or similar instrument to determine which plug causes no change in the operation of the engine when shorted. After the engine has been operating for a short time, a defective plug may often be located by feeling various plugs to determine which one is still cold. Oil which has collected in the cylinder may foul the plugs when a cold engine is first started. This fault usually corrects itself as soon as the engine has been operating for a few minutes.

Moisture during damp weather may cause failure in the ignition system. The ignition system should be thoroughly inspected to determine the presence of any moisture.

The ignition wiring should be examined to see that the insulation is in good condition and that no shorts have developed because of insulation failure. When only one cylinder is misfiring, the trouble is usually located in the spark plugs or the ignition wiring, rather than in a magneto.

The irregular operation of all cylinders or the failure of all cylinders to fire on one set of plugs is usually the fault of an improperly operating magneto. Difficult starting, loud exhaust, overheating of the engine, detonation, and backfiring usually indicate magneto troubles.

When an engine fails to start because of ignition trouble, the cause

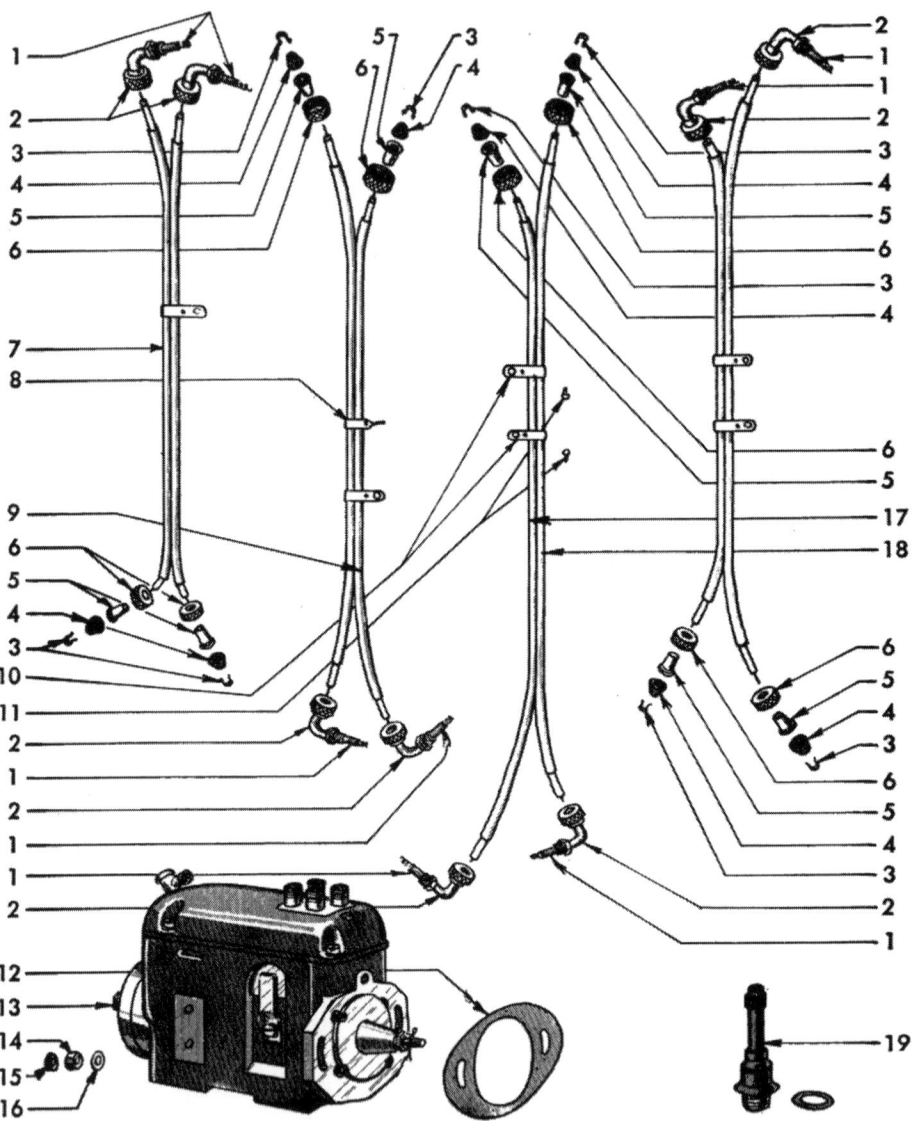

Fig. 24. A radio-shielded ignition system for a light, 4-cylinder, aircraft engine showing (1) terminal sleeve, (2) elbow assembly, (3) magneto terminal clip, (4) rubber gland, (5) cable ferrule, (6) terminal nut, (7) cable assembly, Cyl. 1 and 3 lower, (8) clamp, (9) cable assembly, Cyl. 2 and 4 upper, (10) clamps, (11) fastening screws, (12) magneto flange gasket, (13) magneto, (14) magneto to crankcase cover nut, (15) magneto to crankcase cover palnut, (16) magneto to crankcase cover washer, (17) cable assembly, Cyl. 1 and 3 upper, (18) cable assembly, Cyl. 2 and 4 lower, (19) spark plug. (Courtesy of Continental Motors Corporation)

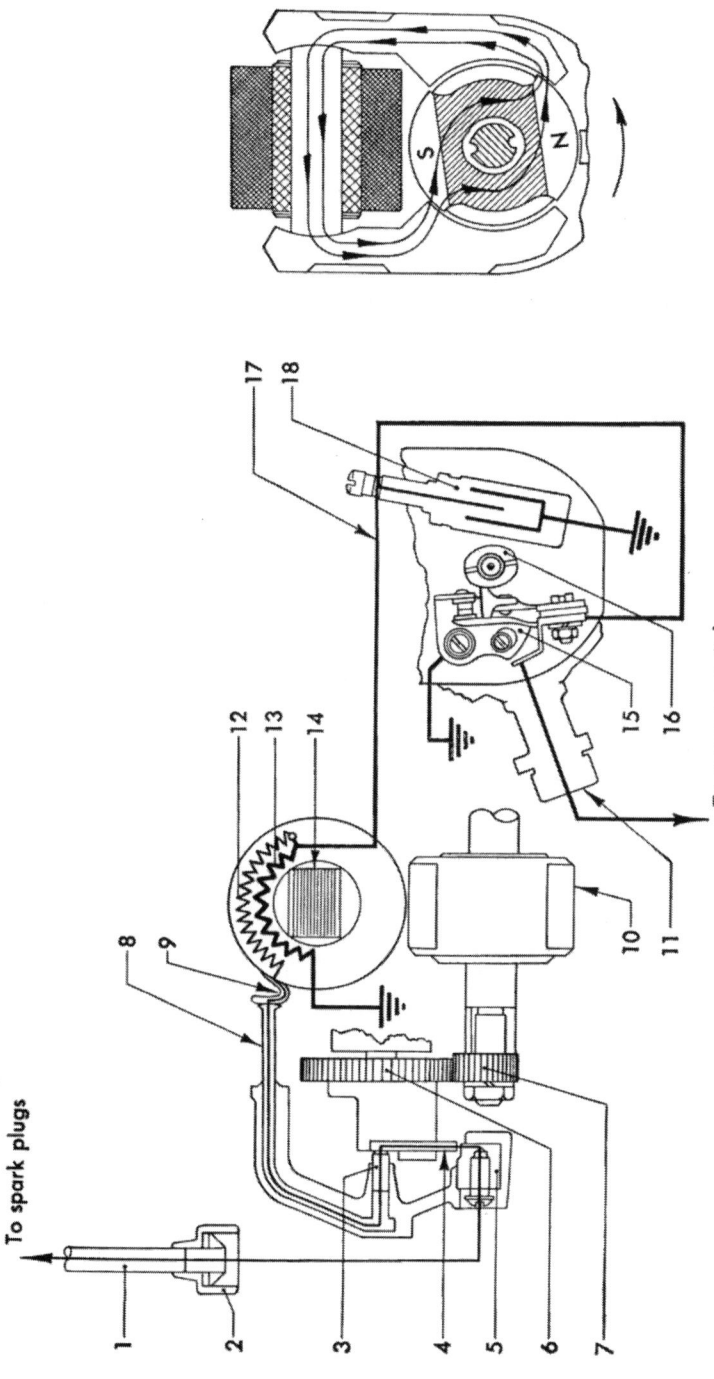

To spark plugs

To magneto switch

S   N

1
2
3
4
5
6
7
8
9
10
11
12
13
14
15
16
17
18

Fig. 25. A schematic diagram of electric and magnetic circuits of an aircraft magneto showing (1) high-tension cable, (2) high-tension terminal, (3) carbon brush and spring, (4) distributor electrode, (5) distributor-plate electrode, (6) distributor gear, (7) pinion gear, (8) spring eyelet conductor, (9) spring clip, (10) magnet rotor shaft, (11) adapter, (12) secondary winding, (13) primary winding, (14) coil core, (15) breaker assembly, (16) breaker cam, (17) winding lead condenser, (18) condenser. (Courtesy Eisemann Corporation)

may be an open or grounded secondary circuit, which includes the secondary wiring, distributor finger, and brush. A grounded or open primary circuit, a shorted condenser, or other faulty wiring will also prevent starting the engine. If the engine operates irregularly on all cylinders, starts with difficulty, and will not develop its full horsepower,

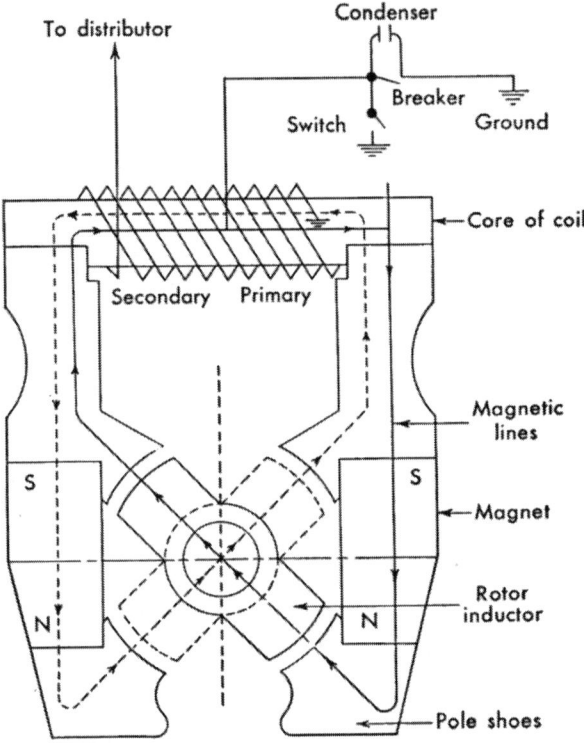

Fig. 26. A schematic drawing showing the fundamental operations of an aircraft magneto.

the cause may be weak magnets or dirty, pitted, or burned contact points. Weak ignition will occasionally cause red exhaust flames. Faulty timing of the ignition may cause detonation, backfiring, low power output, and hard starting. Overheating may occur when the timing is too early. Late timing may cause a low power output, overheating, and loud exhaust. If the engine continues to run after the ignition switch is turned to the OFF position, the magneto ground wires probably have loose connections or have become disconnected. When the ignition switch is in the OFF position, the magneto ground wires are grounded. When the magneto switch is in the ON position, the ground circuit is open.

The ignition system will operate properly if its parts are in proper adjustment and in good condition. Every mechanic should be thoroughly familiar with every part of the ignition system, be able to test every part properly, and should know when it is right.

The testing of magnetos and the ignition timing for each engine

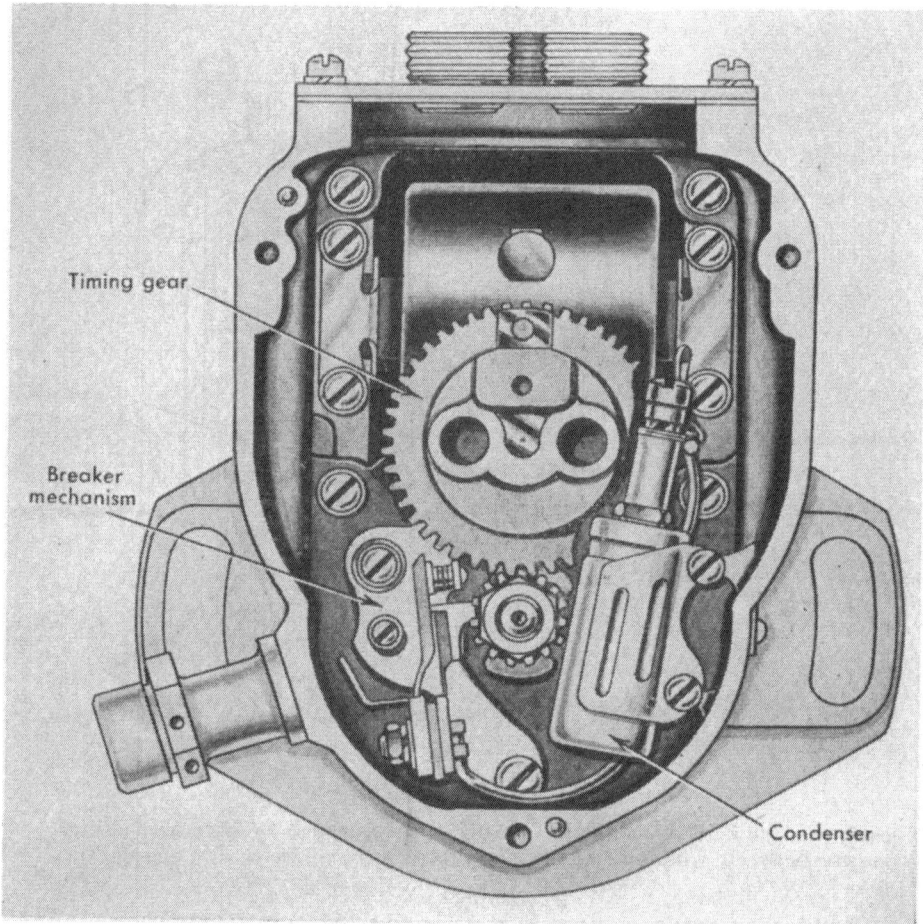

Timing gear

Breaker mechanism

Condenser

Fig. 27. A cutaway view showing the condenser, breaker mechanism, and timing gears of an aircraft magneto. (Courtesy of Eisemann Corporation)

should be done in accordance with the Manufacturer's Manual for the particular engine.

Servicing the Spark Plugs. The servicing and maintenance of the spark plugs are of extreme importance. The spark plug must operate properly in order that the engine may perform its functions and develop its full power. The spark plug must furnish a hot spark to each cylinder every other revolution of the crankshaft in the four-cycle type of engine. This means that when the engine is rotating at 2000 r.p.m., the spark plug must furnish a spark of sufficient intensity that it will ignite the air-fuel mixture 1000 times per minute or approximately 16 times per second in each cylinder.

Spark plugs are usually replaced or completely reconditioned after

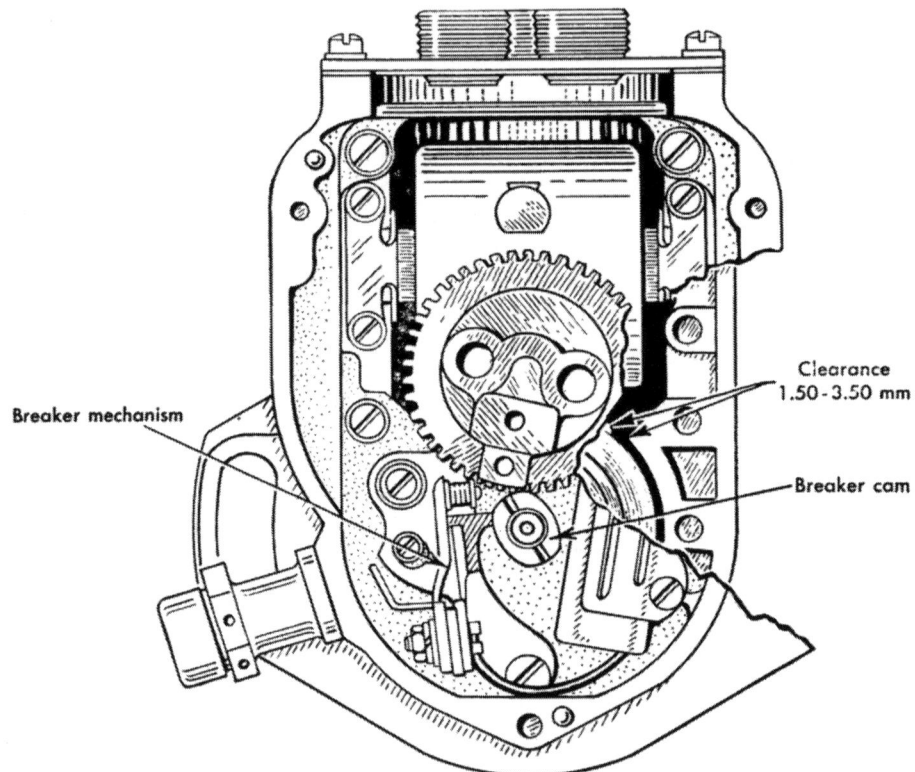

Fig. 28. Drawing of a magneto showing breaker cams, breaker mechanism and clearance between armature and field coils. (Courtesy Eisemann Corporation)

approximately 100 hr. of operation. Cracked or broken insulators on the spark plug and a faulty gap are common causes for the failure of a plug. Excessive oil consumption may cause fouling of the plug, even though the plug itself is in perfect condition. Spark plugs should be cleaned with unleaded gasoline or acetone. Carbon tetrachloride should not be used for cleaning plugs. Mica-insulated plugs should not be cleaned by sandblasting, with steel wool, metal buffing files, or emery cloth, because particles of these materials become embedded in the mica insulator. Any obvious defects in the mica insulator, such as holes, dents, or broken mica, should be cause for rejection of the plug.

The proper tools are essential for efficient service work on spark plugs. Unless the proper tools are used, it is almost impossible to prevent damage to the spark-plug parts. A proper bench should be furnished. This bench should be provided with a moisture-free storage

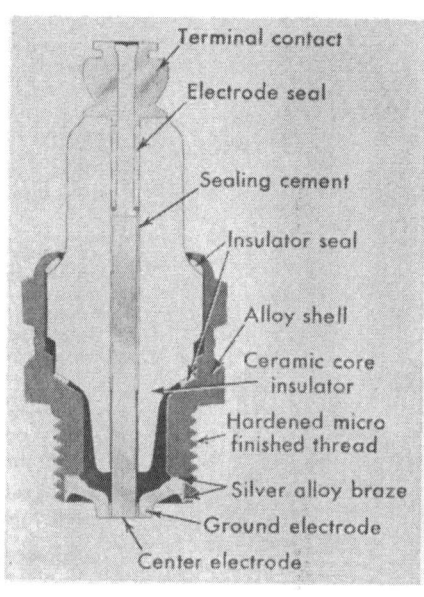

Fig. 29. Short-reach, unshielded, "cold"-type, aircraft-engine spark plug. (Courtesy The Electric Auto-Lite Company)

space for reconditioned plugs. A vise should be provided for the disassembly of the plug.

Servicing Without Disassembly. It is not necessary to disassemble the plugs if they have been functioning properly. The plugs should first be degreased. They should be placed in a wire-mesh basket and lowered into a solution of trichlorethylene which is kept near the boiling point. In order to save the cleaning solution, the tank should be equipped with a condensing coil through which cold water is circulated. This coil should line the tank above the liquid level. Do not inhale the fumes from this chemical. This process will remove all oil and grease, but not the carbon. If only a few plugs are to be cleaned, they may be degreased by scrubbing them in unleaded gasoline with a soft brush. The spark plugs should be left to soak in clean unleaded gasoline.

Removing Carbon. Carbon on the end of the plug can be most efficiently removed by mounting the plug in the proper receiver on the

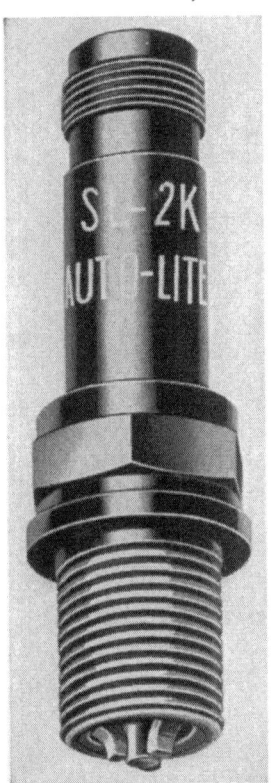
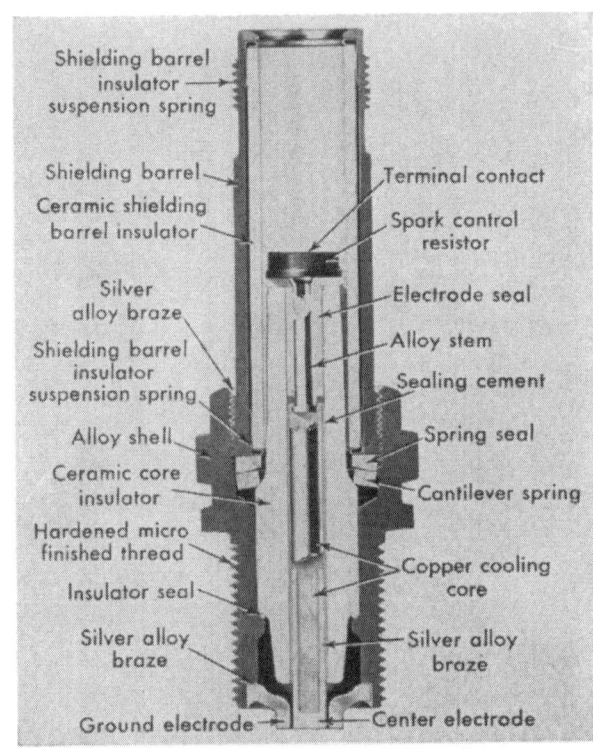

Fig. 30. Long-reach, shielded "hot"-type aircraft-engine spark plug. (Courtesy The Electric Auto-Lite Company)

end of the shaft of a small electric motor. As the plug revolves, the carbon is removed with a soft wire brush. The pressure exerted should be light, and the carbon should be removed within a few seconds.

Checking the Spark Plugs. To check the electrode clearance, which is the spark gap, a wire of the proper gauge recommended by the manufacturer should be used. The wire gauge usually consists of one wire 0.001 in. smaller than the desired gap and the other wire 0.002 in. larger than the desired gap. When the larger wire goes through the gap, or the smaller wire will not go through the gap, the electrode should be reset. The spark gaps vary with different engines and different plugs from about 0.012 in. plus 0.002 in. or minus 0.001 in. up to gaps as great as 0.028 in. or 0.030 in. Irregular or rough operation of the engine is often caused by gaps that are too large or too small.

Adjusting Gaps. The correct setting of the electrodes which determine the size of the gap is important. Particular care must be used at

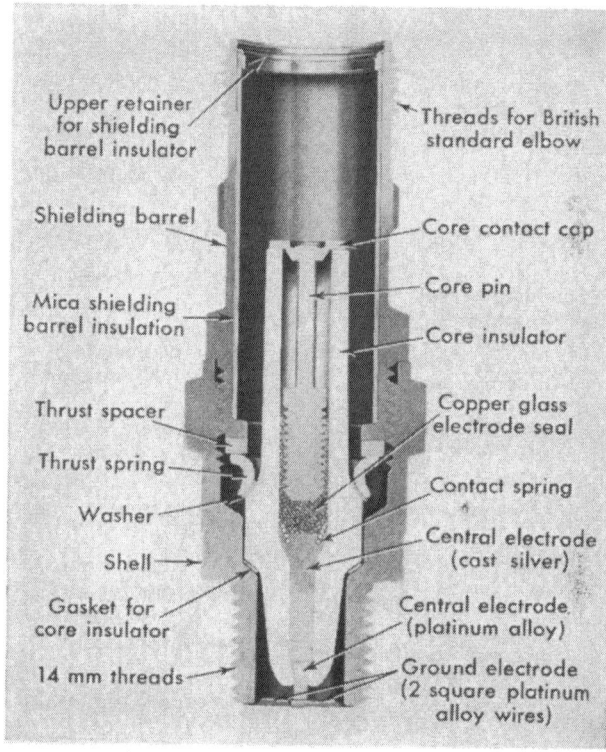

Fig. 31. Shielded aircraft-engine spark plug having two "shell" electrodes. The "shell" electrodes are of platinum. (Courtesy AC Spark Plug Division, General Motors Corporation)

this point in the servicing of the plug. The proper setting of the gaps determines, to a large degree, the number of hours of efficient operation that can be obtained from the plug before servicing is required. Since electrodes vary in shape and adjustment, the proper tools should always be used.

Some plugs have a circular gap while others have from 1 to 4 electrodes attached to the shell. When adjusting the gap, the manufacturer's recommendations for each type of plug should always be followed. After adjusting the gap, it should always be checked with the proper feeler gauge. Unshielded plugs usually may be operated with greater gaps than the shielded type of plug.

Cleaning the Barrel. If the barrel of a shielded spark plug is dirty, it should be cleaned with a soft cloth wound on a stick and dipped in

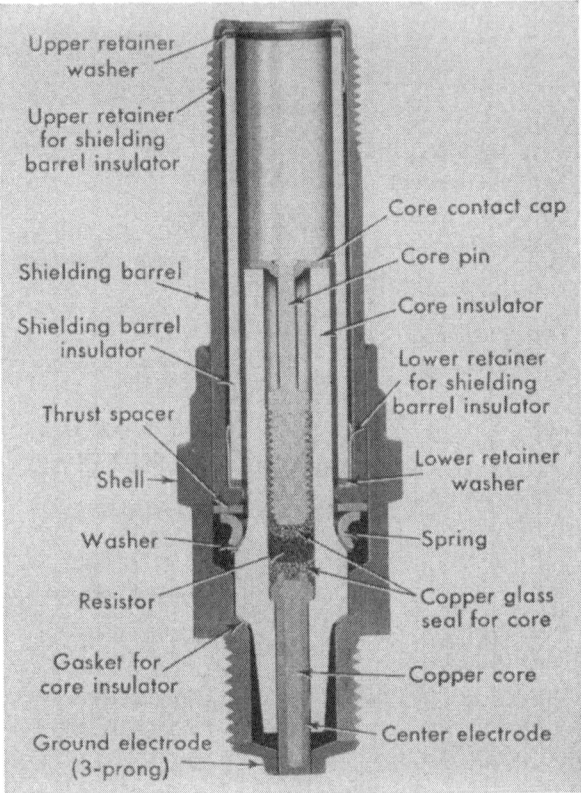

Fig. 32. Shielded aircraft-engine spark plug having three "shell" electrodes. (Courtesy AC Spark Plug Division, General Motors Corporation)

clean unleaded gasoline. If the insulation is of mica, the plug should be wiped in the direction in which the mica is wrapped to prevent damage.

Drying the Plug. In order to dry out any moisture which may have accumulated during the servicing operation, spark plugs should be dried in an oven at a temperature of about 225° F. for 4 hr.

Battery Maintenance. The following are the necessary operations for maintenance of the battery ignition installations:

1. A master battery switch must be connected so that it can *not* disconnect the engine ignition system.

2. The battery should be inspected for proper charge and water level at least every 10 hr. of engine operation and always before flight if the engine has been idle for more than 24 hours. The water evaporates to some extent even though the battery is not in service.

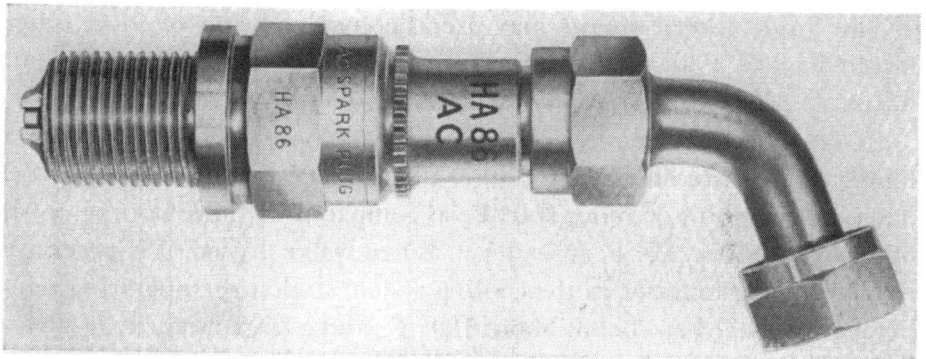

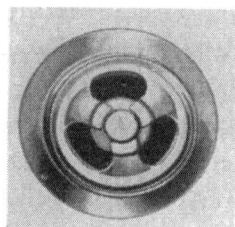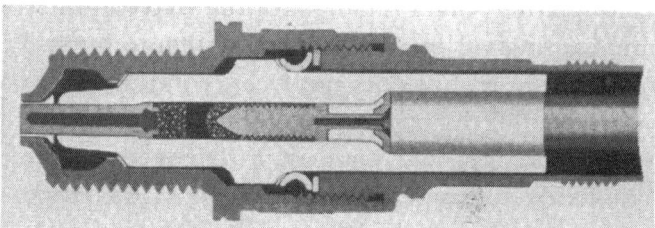

Fig. 33. High-altitude aircraft-engine spark plug. Three views. (Courtesy AC Spark Plug Division, General Motors Corporation)

3. The battery terminal connections and the generator leads should be inspected for tightness every 10 hr.

4. If the battery condition, insofar as charge, is uncertain, the pilot should avoid, if possible, operating auxiliary electrical equipment such as landing gear, flaps, and radio during the take-off period.

4. COOLING AND INDUCTION SYSTEMS

An important part of the power plant is the cooling system. Cowling plays an important part in the cooling of the engine. Whenever aircraft is serviced, the cowling should be carefully inspected to see that it is properly fastened and that there are no cracks developing and that the chafing strips are in good condition.

Engines are cooled both by a liquid coolant and by air. In the air-cooled engine, the air is led about the cylinders, and away by means of baffles, which must be checked at service periods to see that they are properly fastened.

The air-cooled system requires very little service, except inspection to see that there is no accumulation of oil and dirt on the fins and that no fins are cracked or broken.

The liquid-cooled engine may use as a coolant water or some other liquid having a higher boiling point and lower freezing point than water.

In liquid-cooled engines, it is possible to operate the engine at a higher temperature if a coolant other than water is used. Ethylene glycol has a boiling point of about 350° F., as compared with the boiling point of water which is 212° F. at sea level. When water is used, it is necessary to have a large radiator in the cooling system so that the operating temperature may be kept below about 190° F., and a large quantity of water must be carried in the cooling system. Water freezes at 32° F. and, in cold weather, there is danger of the water in the cooling system freezing and causing various parts of the system to burst. Ethylene glycol freezes at about 0° F., but does not freeze solid until a temperature of —48° F. is reached. Ethylene glycol, therefore, protects the cooling system in even extremely cold weather. Ethylene glycol, however, does not circulate freely at temperatures below 0° F. At temperatures below 0° F., it forms a sort of slush. When the airplane is to be exposed to temperatures below 0° F. while stored or standing idle, the ethylene glycol should be drained from the system while hot and heated before it is returned to the cooling system when ready to start the engine. Ethylene glycol has a tendency to loosen rust and scales and should be drained frequently and strained before returning to the system.

If water is used in the cooling system during cold weather, it should be diluted with some such substance as ethylene glycol, glycerin, or alcohol to prevent freezing. Ethylene glycol is preferred as it dissolves in water in any proportion and freezes in the form of a slush instead of in the form of a solid which would cause damage to the system. Ethylene glycol expands when heated to a greater extent than water, and a proper amount of expansion space in the cooling system should be provided. When filling a system with ethylene glycol, the system is filled to the ordinary water level and then approximately 1 gal. of the solution is drained out. In air-cooled engines, a high degree of cooling efficiency is maintained by a streamlined, ring-type cowling around the outside circumference of the cylinders on radial engines. One problem with which the engineer is confronted in cooling is that, when the engine is cooled efficiently at low speeds, it may not be cooled efficiently at high speeds.

Fig. 34. A cutaway view of a 5-cylinder, radial, aircraft engine showing the accessory drive gears and induction system. (Courtesy Jacobs Aircraft Engine Company)

If the cooling system is designed to cool the engine efficiently at high speeds, it may not be effective at low speeds.

Oil plays an important part in cooling the engine. Engine oil is circulated through all the internal parts of the engine. On its way through the engine, it becomes heated. After the oil leaves the engine, it is often circulated through an oil radiator to cool it before it is returned to the supply tank. On light engines which are not equipped with oil coolers, the hot oil is returned to the supply tank where it is cooled by mixing with the cool oil in the tank. The oil is also cooled as it passes through the oil lines.

Liquid-cooling systems are subject to leaks and stoppages due to the formation of scale and rust, and the depositing of solid particles from the liquid. It is necessary that these systems be drained and flushed at regular intervals. The air-cooling systems are not subject to such troubles as these, but it is necessary that the cowling flaps be in proper operating condition and that all baffles be checked at frequent intervals to see that they are securely fastened and safetied. Oil coolers should be checked at regular intervals to see that they have not become clogged and that the controls are in good operating condition.

III GLOSSARY OF TERMS USED IN AIRCRAFT ENGINES

acceleration. Change in velocity.

aircraft. Any weight-carrying device designed to be supported by the air, either by buoyancy or by dynamic action.

airfoil. Any surface, such as an airplane wing, aileron, or rudder, designed to obtain a reaction from the air through which it moves.

airplane. A mechanically driven fixed-wing aircraft, heavier than air, which is supported by the dynamic reactions of the air against its wings.

alloy. A metal made up of two or more metals usually formed by melting them together.

alternating current (ac.). Electric current which flows in one direction, then in the opposite direction, reversing at regular intervals.

alternator. A machine for generating alternating current. An alternating-current generator.

altitude. Height above some given level. Ceiling is given in feet above ground; altimeters generally give height above sea level.

ampere. The unit for measuring the rate of flow (current) of electricity.

aneroid. A metallic cell from which the air has been partially exhausted, arranged to contract or expand with changes in gaseous pressure.

annealing. Heating a metal to a temperature above its critical temperature and then allowing it to cool slowly. This usually softens the metal.

anode. A positive or plus terminal of a battery or other electrical source.

antiknock value (antidetonation). The octane rating of a fuel.

arcing. The jumping of an electric current from one conductor to another through the air.

armature. The revolving core of an electric generator or magneto.

articulated rod (link rod). One of the connecting rods of a radial engine which is attached to the master rod by means of knuckle pins.

atmospheric pressure. The normal pressure of the atmosphere at any given elevation.

autosyn. One of the remote indicating systems.

axial motion. Motion along the axis of a revolving body.

baffle, cylinder. Sheet-metal deflectors used to direct air around cylinders to ensure proper cooling.

battery. A cell or device for supplying an electric current by chemical means.

battery, storage. A cell which is restored by driving an electric current backward through the electrolyte.

British thermal unit (Btu). A unit of heat measure. The amount of heat necessary to raise the temperature of 1 lb. of water 1° F.

cam. A projecting part on a shaft, rim, plate, or wheel to impart a desired movement to a follower in contact with that part of the shaft, rim, plate, or wheel.

camber. The convex or concave curvature of the surface of an airfoil.

cam plate. A flat circular plate having cam lobes around its circumference.

camshaft. A shaft which carries cam lobes.

carburetor. A device for measuring and mixing the proper amount of fuel with air to form the air-fuel mixture.

cathode. The negative electrode of a source of electric current.

cell, dry. An electric cell having all of its parts sealed against the atmosphere. The materials in this cell are not dry.

cell, wet. An electric cell in which the electrolyte is entirely liquid and not sealed from the atmosphere.

centrifugal force. The tendency of a body to move in a straight line when forced to move in a curved path.

centrifugal pump. A pump which makes use of centrifugal force in its operation.

combustion chamber (compression chamber). The total space within the cylinder and cylinder head when the piston is at top dead center.

compression ratio. The ratio between the space within the cylinder and combustion chamber when the piston is at bottom dead center, and the space within the cylinder and combustion chamber when the piston is at top dead center.

congeal. An increase in viscosity of a liquid, usually due to change in temperature.

coulomb. A unit used to measure quantities of electricity.

counterweight. A weight attached to a revolving part such as a crankshaft to bring about conditions of balance.

crankcase. A part of an internal-combustion engine that encloses the crankshaft and connecting rods.

crankpin (wrist pin). The part of the crankshaft about which the connecting-rod bearing fits.

crankshaft. The main rotating shaft of an engine by means of which the power developed in the cylinders is changed to rotary motion.

cycle. One of a complete series of recurring events.
   four-cycle principle (Otto cycle)
   One power stroke is developed for

each four strokes of the piston in the cylinder.

two-cycle principle. One power stroke is developed for each two strokes of the piston in the cylinder.

cylinder. A chamber in an engine in which a piston slides back and forth.

Diesel engine. An internal-combustion engine in which the air-fuel mixture is ignited by heat of compression within the cylinder.

density. Actual weight per unit volume of any substance.

detonation. In an engine, a rapid combustion replacing normal combustion. When detonation occurs, there may be loss of power, engine overheating, and noise.

diffuser. A device for reducing the speed of the air-fuel mixture leaving the supercharger impeller and thereby permitting a charge of uniform density to enter each cylinder.

direct current (dc). Current that flows in one direction.

distill. A process by which a liquid is changed to vapor and then condensed back to a liquid.

dynamic damper. A short pendulum in the form of a counterweight, designed to damp out engine vibrations set up by the power strokes.

dynamotor. Combination of generator and motor.

electrode. Either terminal of an electric source. An electrode may be a wire, plate, or other conducting object.

electron. A negative particle of electricity.

element. A simple substance made up of only one kind of atom, such as iron, aluminum, or oxygen.

engine, internal-combustion. Any engine in which the energy is produced in the engine cylinder.

engine, external-combustion. A heat engine which derives its heat from fuel consumed outside the engine cylinder.

exhaust port. The opening from which gases escape from the cylinder after combustion.

exhaust valve. A valve for opening and closing the port through which the exhaust gases escape from the cylinder.

feathering. Rotating a propeller blade so that the leading edge meets the air in such a manner that the air stream is approximately parallel to the chord of the blade.

flywheel. A wheel on an engine crankshaft to counteract variable pressure during the strokes of the piston and to carry the engine over the dead centers.

foot pound (ft.-lb.). A unit of energy, or work, being equal to the work done in raising 1 lb. avoirdupois against the force of gravity a distance of 1 ft.

forging. Metal fabricated by the blows of a hammer or a die forcing the material to conform to the contours of the forming die.

fuel. Any substance used to produce heat by burning.

fuselage. The main structure of an airplane, approximately streamlined in form, to which are attached the wings and tail units.

generators. A general term applied to machines which are used for the transformation of mechanical energy into electrical energy.

glide. The normal flight of an airplane without power.

governor. A regulating device.

graduations. Marks setting off intervals such as inches or degrees on a measuring instrument.

gyroscope. A rapidly revolving, heavy wheel.

gyroscopic. Pertaining to an action brought about by the use of a gyroscope.

horsepower. That power necessary to raise 1 lb. 33,000 ft. in 1 min., or that power which will raise 33,000 lb. 1 ft. in 1 min. against the pull of gravity.

hydraulic. Pertaining to liquids in motion.

hydrometer. An instrument for measuring the density of liquids.

idling. The speed of rotation of an engine with the throttle closed.

ignition. The act of igniting; subjection to the action of fire or intense heat.

induced current. Current set up in a conductor due to the effect of a magnetic field.

induction coil. An apparatus for transforming an ordinary battery current by induction into an alternating current of high potential.

impeller. The vaned rotating part of a supercharger.

inertia. The tendency of a body to continue in its state of rest or motion.

inflammable. Burns readily.

intake valve. A valve for opening and closing the port through which the charge of air-fuel mixture enters the cylinder.

jet propulsion. Propulsion by means of the reaction of gases under high pressure escaping through an opening.

jig. A pattern, form, or framework, accurately dimensioned and aligned, in which identical structures of parts can be produced to meet a standard.

lapping. Finishing a metal surface to a high polish by the use of a fine abrasive.

linear. Lengthwise.

link rod. (*See* articulated rod.)

loading, power. The gross weight of an airplane divided by the rated horsepower of the engine computed for air of standard density, unless otherwise stated.

loadstone. A naturally occurring iron ore which attracts magnetic substances.

lobe. A rounded projection of a cam; it usually refers to the projections on an ignition timer-distributor shaft for operating the breaker contact points.

lubricant. A substance used to reduce friction.

magnetic field. Any space through which magnetic influence is exerted.

magneto. A device for producing electricity dependent on permanent magnets for its magnetic field; it is used for ignition in internal-combustion engines.

main bearing. A bearing supporting the crankshaft of an internal-combustion engine.

manifold. A pipe or casting with several outlets, used between the carburetor and cylinders of an internal-combustion engine or to carry off the exhaust heat and gases.

meshes. Fits together; for example, two gears.

metallurgy. The science of metals.

metering jet. An opening of predetermined size which regulates the flow of fuel through the carburetor.

molten. In liquid form.

momentum. The energy contained in a body due to its motion.

nacelle. An enclosed shelter for personnel or for a power plant. A nacelle is usually shorter than a fuselage and does not carry the tail unit.

neutron. An uncharged particle in the nucleuses of atoms.

octane. The antiknock rating given in numbers.

octane rating. The percentage of iso-octane in a mixture of iso-octane and normal heptane required to match the performance of the fuel being tested in a special test engine under controlled conditions.

ohm. The practical unit of electrical resistance, being the resistance of a circuit in which a potential difference of 1 volt produces a current of 1 ampere.

piston. The plunger which moves within the cylinder of an engine or pump. The efficiency of compression depends largely on the proper fitting of the piston.

precision instrument. An instrument for accurately measuring extremely small quantities.

preignition. The ignition of the compressed gas in the combustion chamber before it is desired, due to a heated carbon deposit or other hot surface.

propeller. Any device for propelling a craft through a fluid, such as water or air; especially a device having blades which, when mounted on a power-driven shaft, produce a thrust by their action on the fluid.

adjustable propeller. A propeller whose blades are so attached to the hub that the pitch may be changed while the propeller is at rest.

automatic propeller. A propeller whose blades are attached to a mechanism that automatically sets them at their optimum pitch for various flight conditions.

controllable propeller. A propeller whose blades are so mounted that the pitch may be changed while the propeller is rotating.

geared propeller. A propeller which rotates at a speed different from that of the engine crankshaft due to a train of gears between the crankshaft and the propeller shaft.

pusher propeller. A propeller mounted on the rear end of the engine or propeller shaft.

tractor propeller. A propeller mounted on the forward end of the engine or propeller shaft.

propeller area. The blade area times the number of blades.

propeller efficiency. The ratio of the thrust power to the input power of a propeller.

propeller pitch. The distance the propeller would screw forward in a semisolid due to the angle of blade setting.

effective pitch. The distance an aircraft advances along its flight path for one revolution of the propeller.

static pitch. A condition where the propeller rotates in the same path without forward motion, as when warming up the engine.

theoretical pitch. The distance a propeller would move through the air if no slip occurred.

propeller radius. The distance of the outermost point of a propeller blade from the axis of rotation.

propeller slip. The difference between theoretical pitch and effective pitch.

propeller thrust. The component of the total air force on the propeller which is parallel to the direction of advance.

propeller tipping. A protective covering of the blade of a propeller near the tip.

proton. A positive charge of electricity.

p.s.i. Pounds per square inch.

pulsating direct current. Current which flows in one direction all the time, but is periodically interrupted or changes its intensity at more or less regular intervals.

ram effect. The pressure built up by the velocity of a fluid such as air.

rarefied atmosphere. Atmosphere of less density than standard.

resistor. An electrical resistance such as a resistance coil.

rheostat. A variable resistance. Usually, a radial arm touches a circular resistance so that, when it is rotated, more or less of the resistance is included in the circuit.

rocker arm. An oscillating arm borne by a shaft, which is usually used to operate valves.

rocket. A device propelled by the burning of fuel contained within the device itself, for example, a sky rocket.

S.A.E. (Society of Automotive Engineers) *or Saybolt numbers.* These numbers are used to indicate the viscosity of a liquid such as oil.

safetied. Fastened securely in place.

scavenger oil (or **scavenged oil**). Used oil which is removed from an engine after it has passed through the bearings, etc. Oil removed from a dry sump engine by the scavenger pump.

scavenger pump. A pump for removing used oil from the engine and returning it to the oil tank.

servo unit. A cylinder containing a piston which moves back and forth bringing about a desired movement such as the moving of an aileron by the action of the automatic pilot.

shielding (radio). A method by which radio interference due to the aircraft electrical system is eliminated.

sodium. A metallic element.

solenoid. Operated by means of a magnetic coil which is energized by means of an electric current causing a soft iron core to move due to magnetic attraction.

spark plug. A plug which is used to ignite the air-fuel mixture.

specific gravity. The number of times the weight of a unit volume of a substance is of the weight of an equal volume of water.

spinner. A fairing of approximately conical or parabolical shape which is fitted coaxially with the propeller hub and revolves with the propeller.

splines. A series of lands and grooves cut in a shaft to prevent its turning when inserted into matching lands and grooves.

stellite. A nonferrous, cobalt-chromium-tungsten alloy used for hard-surfacing other metals or alloys.

sump, dry. An engine in which the oil supply is carried other than in the crankcase of the engine.

sump, wet. An engine in which the oil supply is carried in the crankcase.

supercharger. A device, usually a rotating fan, for supplying the engine with a greater weight of air-fuel mixture than would normally be inducted at the prevailing atmospheric pressure.

surging. Irregular flow of a liquid.

synchronize. To cause to agree in time, as to time a magneto armature with the distributor so they agree in their timing; to cause an agreement in the timing of two separate ignition systems attached to one engine.

tachometer. An instrument that indicates in revolutions per minute the rate at which the crankshaft of an engine turns.

tappet. A mechanism between the cam and valve or push rod.

tension. A pulling force.

thermal circuit breaker. A circuit breaker operated by means of changes in temperature.

torque. That force which produces or tends to produce rotation.

torsional force. A twisting force.

transformer. Two coils of wire wound on a common iron core used to change the voltage of an electric current by induction.

vacuum. A true vacuum is a space entirely empty of all matter, even traces of gas.

valve. A device for opening or closing a passageway or a port.

by-pass relief valve. A valve designed to regulate the oil or fuel pressure within the engine by allowing excess oil or fuel to by-pass the pump.

valve guide. A bearing through which the valve stem passes.

valve tappet. A tappet between a cam and a valve.

vapor pressure. The pressure exerted by a vapor within a confined space.

Venturi tube. A short tube of varying cross section. The flow through the Venturi causes a pressure drop in the smallest section, the amount of the drop being a function of the velocity of the flow.

volatility. A property of a liquid which determines the temperature at which it passes into the vapor state. Too high volatility in a fuel may produce vapor lock. Too low volatility produces difficulty in engine starting at low temperatures.

volt. That electromotive force which if steadily applied to a conductor whose resistance is 1 ohm will produce a current of 1 amp.

voltage booster. A device for increasing the voltage of an electric current.

volumetric efficiency. The highest possible filling of a combustion chamber and cylinder with air-fuel mixture when the piston is at bottom dead center.

watt. A unit of electric power found by multiplying amperes by volts.

Wheatstone bridge. A device for the measurement of resistances.

windmilling. The turning of a propeller by the force of the air stream.

wobble pump. A hand fuel pump used for starting or for emergency purposes. The handle may be provided with holes by means of which an extension rod may be attached for operating the pump at some distance from the pilot's seat.

wrist pin. (*See* crankpin.)

IV INSPECTION PROCEDURES

The following does not intend to give complete detailed inspection procedures for any individual power plant, but may serve as a guide in making power-plant inspections. When inspecting a power plant, the manufacturer's manual should be followed whenever possible. Power plants vary greatly in size, complexity, and arrangement of the various parts. They are, however, fundamentally the same, and the following general instructions may apply to any or all of them.

Engine Cooling Systems. Daily or before flight, cooling systems should be checked for proper coolant level and proper fastening of filler caps.

After every 25 hr., the cowling flaps and oil-cooler shutters should be carefully checked and lubricated. The coolant pump should be inspected for leaks, and the pump-shaft bearings should be lubricated. If necessary, the packing should be adjusted or replaced. Radiators, expansion tanks, lines, and connections should be inspected for leaks. Hose clamps should be checked for tightness, drain cocks for proper safetying, and shutters for proper functioning.

At the 50-hr. period, the cooling system should be drained and flushed. Drain plugs should be checked for tightness and safety. The ball-check valves should be removed from the filler unit on the expansion tanks, and the ball inspected for freedom of movement. After draining ethylene glycol from the system, it should be flushed with boiling hot water. The ethylene glycol should be strained, and its specific gravity should be checked before it is returned to the engine. The temperature indicator should be checked for accuracy.

If the engine is air cooled, the cooling fins should be checked to see that they are not being rubbed by the cowling and that none is broken. This should be done daily or before flight. At the 25-hr. period, all cylinders should be checked for damaged or broken fins and to see that baffles are in place and properly safetied. Check the magneto and cool-

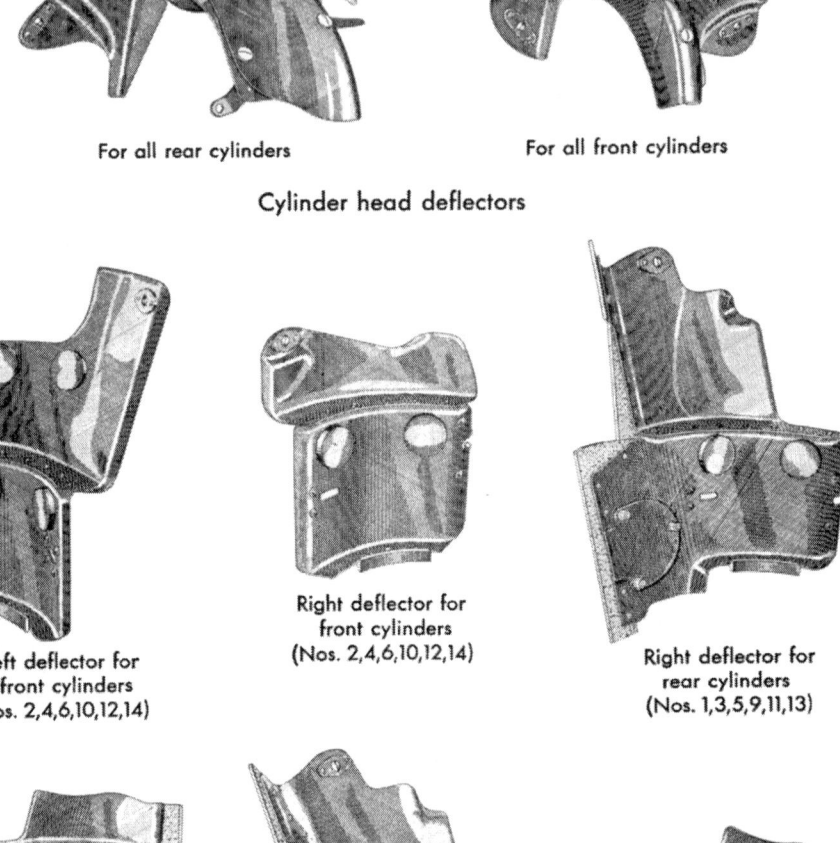

For all rear cylinders

For all front cylinders

Cylinder head deflectors

Left deflector for
front cylinders
(Nos. 2,4,6,10,12,14)

Right deflector for
front cylinders
(Nos. 2,4,6,10,12,14)

Right deflector for
rear cylinders
(Nos. 1,3,5,9,11,13)

Left deflector for
rear cylinders
(Nos. 1,3,5,11,13)

Right and left deflector
for rear cylinder No. 7

Left deflector for
rear cylinder No. 9

Fig. 35. Cylinder. air deflectors or baffles for a twin-row, 14-cylinder, radial aircraft engine. (Courtesy Wright Aeronautical Corporation)

ing tubes for cleanliness, breaks, cracks, or indications of looseness. The entire engine system and cowl flaps should be checked for excessive wear and proper adjustment.

Fig. 36. A radial engine with air deflectors or baffles between the cylinders. (Courtesy Jacobs Aircraft Engine Company)

Valves. At the 50-hr. period, the valves should be inspected and adjusted. The push-rod hose clamps and tappet guides should be checked for tightness. Each 200 hr. all engines having automatic valve lubrication should have the valves inspected and adjusted.

Engine Controls. Daily or before flight, all engine controls should be checked for proper operation, looseness, slack, proper safetying, and general condition. On engines equipped with superchargers and propeller controls, these items should be checked at the 25-hr. period. All control assemblies, such as rods, cables, guides, pulleys, and linkage supporting brackets, should be carefully checked for free and full movement. A careful check should be made for lost motion, bent rods, frayed cables, broken, loose, misaligned or damaged pulleys, and

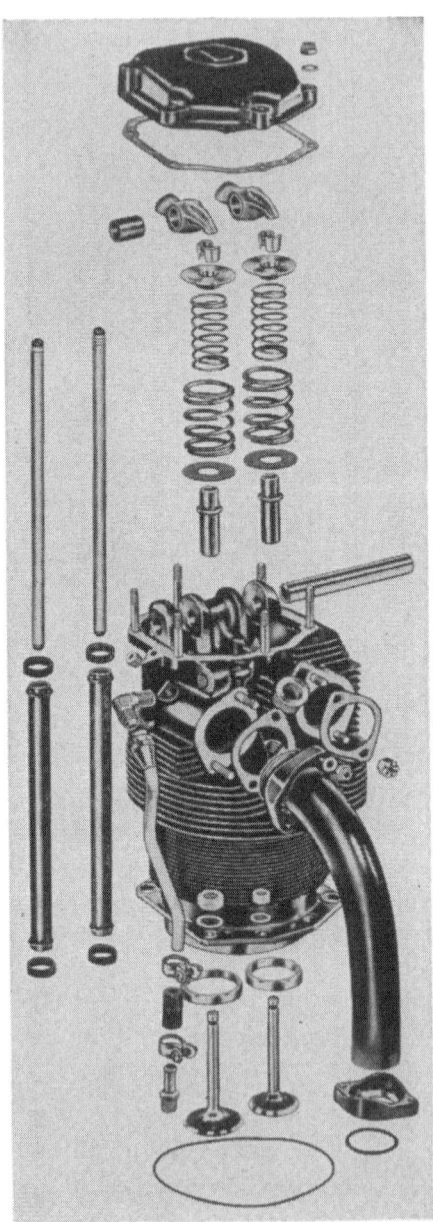

Fig. 37. An exploded view of a cylinder and its accessories. (Courtesy Lycoming Division, The Aviation Corporation)

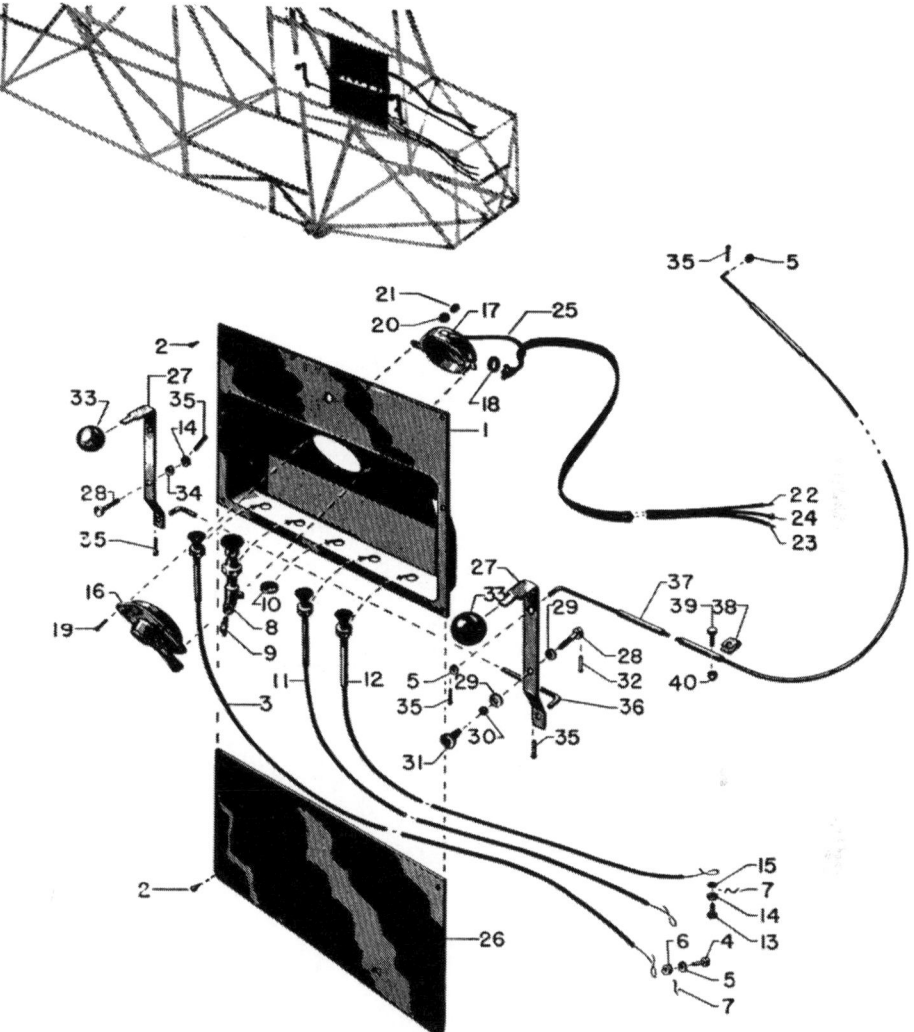

Fig. 38. Power plant controls and control panel showing (1) control panel, (2) stove head screw, (3) cabin heater control, (4) bolt, (5) plain washer, (6) shear nut, (7) safety wire, (8) primer pump, (9) male inverted elbow, (10) plug button, (11) carburetor heater control, (12) mixture control, (13) bolt, (14) shear nut, (15) plain washer, (16) dual ignition switch, (17) ignition switch shield assembly, (18) elastic grommet, (19) roundhead machine screw, (20) lock washer, (21) hex machine screw nut, (22) dual ignition right hand shielded hot wire assembly, (23) dual ignition lefthand shielded hot wire assembly, (24) wire assembly, ignition switch shielded ground, (25) bonding braid, (26) plate and control panel accessory cover, (27) throttle lever, (28) aircraft bolt, (29) throttle lever washer, (30) lock washer, (31) throttle tightening knob, (32) cotter pin, (33) throttle knob, (34) burr washer, (35) cotter pin, (36) throttle connecting rod, (37) throttle control cable, (38) flexible throttle attaching clevis, (39) aircraft bolt, (40) self-locking nut. (Courtesy Taylorcraft Aviation Corporation)

loose or missing bolts; and these items should be marked for immediate repair.

Each 25 hr. throttle shafts and bearings should be lubricated with light oil. At the 50-hr. period, fire-wall guides or fair-leads should be lubricated with grease.

The controls which must be checked should include the throttle,

Fig. 39. A typical installation of a five-cylinder radial aircraft engine. (Courtesy Kinner Motors, Inc.)

shutter, carburetor, air-heater, supercharger, air-fuel-mixture, and propeller controls. Sealed bearings should not be lubricated. All controls should be adjusted to move freely, but they should be tight enough to prevent creeping. All moving connections and bell cranks should be cleaned and lubricated with the proper grade of oil each 50 hrs.

Engine. If the engine is mounted in a dynamic engine mount, all fastenings should be carefully checked and lubricated in accordance with the manufacturer's directions. Rocker boxes should be inspected for tightness and oil leaks. The compression on all cylinders should be checked daily or before flight. This is done by turning the engine over by hand four or five revolutions. This operation should always be performed to check for accumulation of oil in the lower cylinders. This should be done without fail if an engine has stood idle for two or three hours or more. The entire engine should be inspected for evidence of oil leaks. The engine should be "run up" on the ground to check the proper functioning of all engine instruments and controls. At the 25-hr. period all cylinder hold-down and coolant jacket fastenings should be

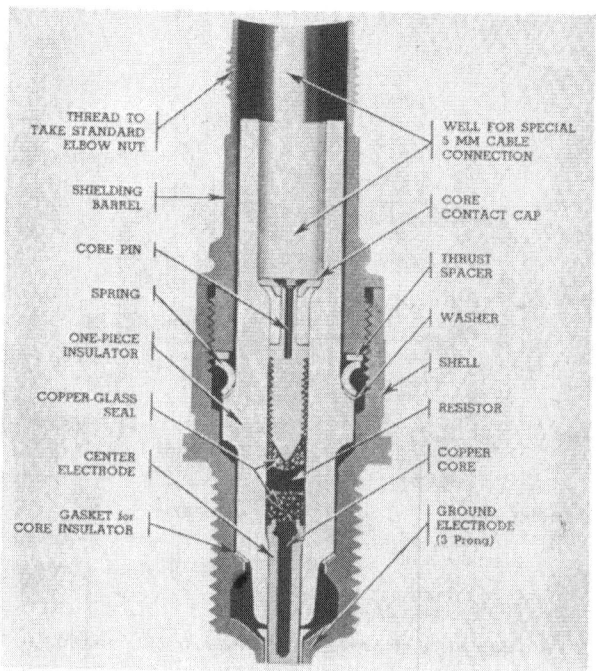

THREAD TO
TAKE STANDARD
ELBOW NUT

SHIELDING
BARREL

CORE PIN

SPRING

ONE-PIECE
INSULATOR

COPPER-GLASS
SEAL

CENTER
ELECTRODE

GASKET for
CORE INSULATOR

WELL FOR SPECIAL
5 MM CABLE
CONNECTION

CORE
CONTACT CAP

THRUST
SPACER

WASHER

SHELL

RESISTOR

COPPER
CORE

GROUND
ELECTRODE
(3 Prong)

Fig. 40. High altitude aircraft engine spark plug. (Courtesy AC Spark Plug Division, General Motors Corporation)

inspected for proper tightness and safetying. The entire engine should be carefully inspected for missing nuts, bolts, broken studs, cracks, broken parts, leakage, and proper safetying. The breather screens should be inspected and cleaned, if necessary.

Electrical System. Daily or before flight, all switches and magnetos should be checked for proper operation and functioning. At the 25-hr. period, spark plugs should be tested for proper shielding to determine any looseness or poor connections, and if the engine is to be operated at high altitudes, spark-gap clearances should be tested and adjusted.

At the 50-hr. period, the magnetos should have the breaker cover removed. The breaker mechanism should be carefully inspected for damaged breaker felts or cushions, damaged cam followers, and weak or broken breaker-arm springs. The springs should be tested for proper tension by the use of a scale. Proper clearance should be determined. The whole assembly should be thoroughly cleaned and checked for any looseness, slack, security of mounting, and proper clearances. The magneto should be lubricated as required. The distributor should be inspected for a cracked rotor or head, sticking or broken brushes, tight-

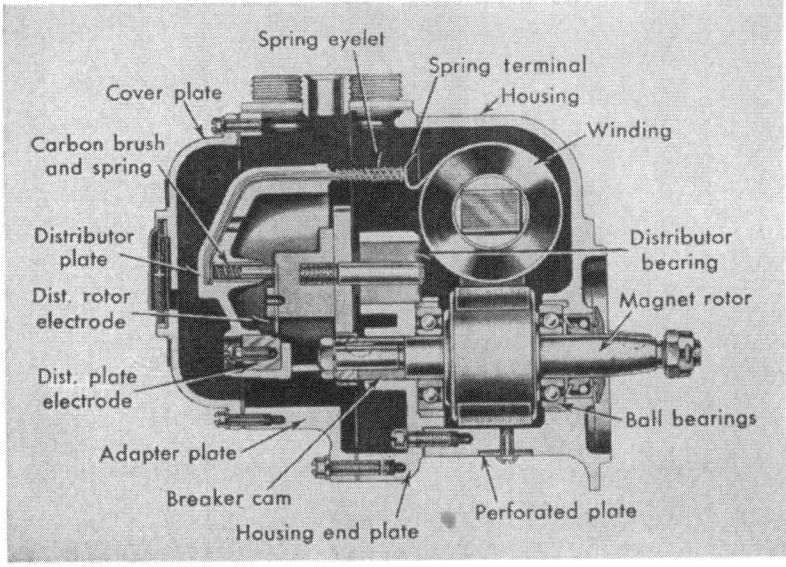

Fig. 41. A cutaway view showing construction of a magneto. (Courtesy Eisemann Corporation)

ness of mounting screws, signs of arcing and for tightness and cleanliness. The ignition cables should be carefully checked for the condition of terminals, connections, exposed ends, or damage to insulation because of chafing.

The ignition boosters should be checked for security of mounting and proper adjustment. All electric leads, wires, cables, and terminals should be checked for security of fastening. With the engine running at about one-third throttle, the ignition switches should be checked in their various positions. If the engine does not stop firing at once when the ignition switch is turned to the OFF position, check the switch and ground connections for defects or looseness.

This test should not be made with an excessively hot engine. It is not necessary to leave the switch in the OFF position until the engine has been completely stopped to make this test. If the engine continues to fire with the ignition switch in the OFF position, it is necessary to stop the engine by turning off the fuel. When this condition exists, care should be taken not to touch the propeller until the fault has been found and corrected.

Generator. Daily or before flight, the generator should be checked for the proper condition. The voltmeter, and the ammeter should be

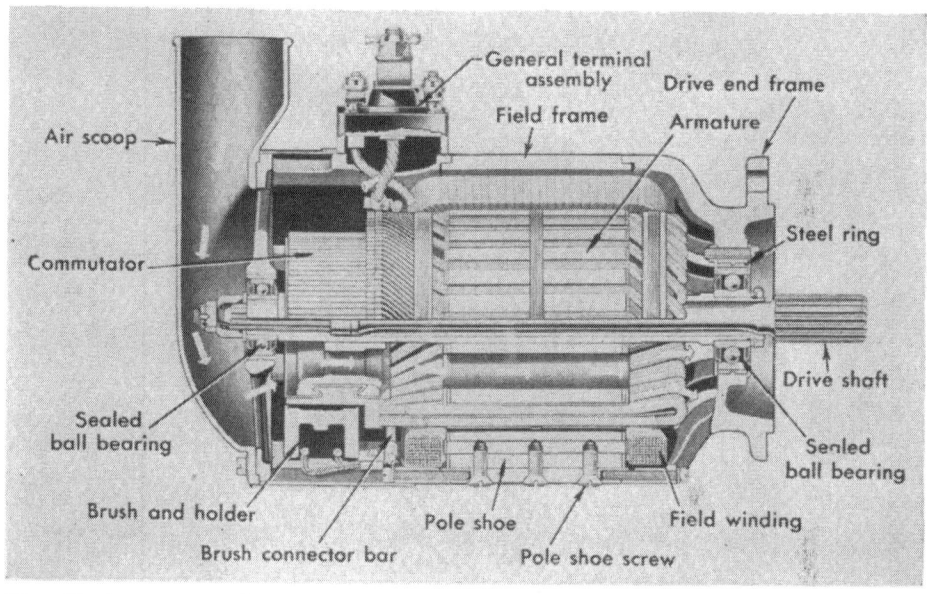

Fig. 42. A cutaway view of an aircraft generator showing construction and parts. (Courtesy Delco-Remy Division, General Motors Corporation)

checked for the zeroing of pointers, cracked cover glasses, and the condition of the electric leads to these instruments. The generator should be inspected for oil leakage and, during the run-up period, the operation of the voltmeter should be checked to determine whether the voltmeter needle is steady and the normal voltage is being generated. The ammeter pointer should indicate ZERO when the main-line switch is in the OFF position. The ammeter pointer should be steady and indicate a normal charge when the main-line switch is in the ON position and the engine is running at normal cruising r.p.m. After high-altitude flights, the generator brushes should be checked for wear.

At the 50-hr. period, the generator should be checked for proper safetying, arcing, security of mounting, and general cleanliness. The generator brush assembly should be checked for worn or sticking brushes, proper spring tension, and loose connections. The commutator should be checked for pits, worn or out-of-round condition, roughness or indication of excessive arcing. All generator leads, connections, and terminals should be checked for general condition and tightness.

The dials on the voltmeter and ammeter should be examined for dis-

Fig. 43. An electric direct cranking starter. (Courtesy Eclipse-Pioneer Division, Bendix Aviation Corporation)

colored or chipped luminous markings. The generator-control panels should be inspected for security of mounting, proper safetying, and cleanliness. The condition of all leads, connections, and terminals should be inspected for tightness and proper fastening, as well as for the condition of the insulation. The voltage regulator, reverse-current cutout, and current limiter should be checked for cleanliness, condition of leads, points, connections, security of mounting, and proper operation.

Retracting Motors. Approximately every 50 hr., the retracting motors should be checked for security of mounting, dirty or rough commutators, proper condition of brush assembly, loose connections, proper tension of brush springs, general condition of housing, mounting, safetying, tightness of housing bolts, general cleanliness, and proper lubrication.

Starters. Daily or before flight, the starters should be inspected for any indication of oil leakage. If any indication of excessive oil is found, the starter should be replaced or put in the proper condition before

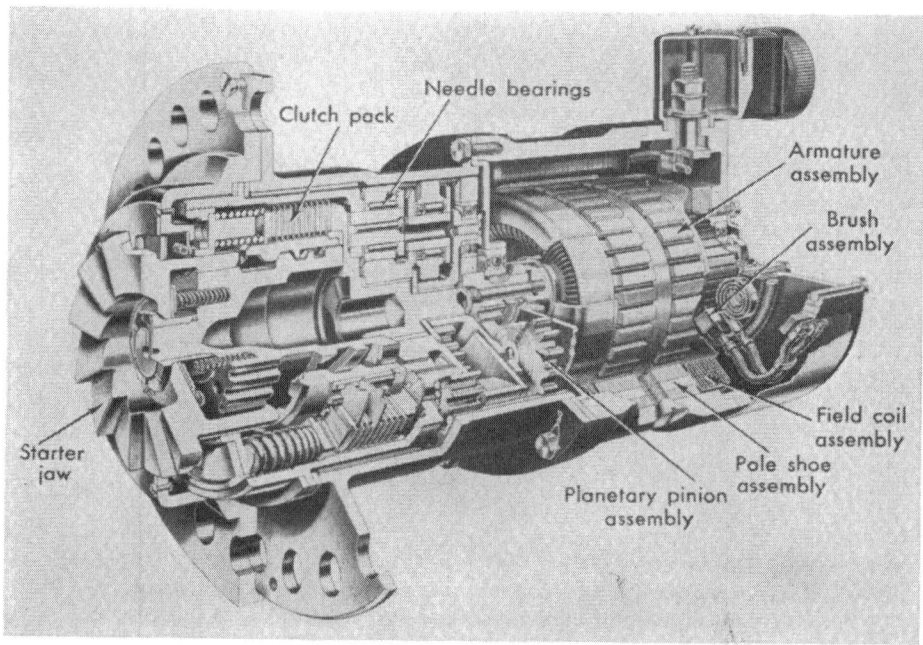

Fig. 44. A cutaway view of an electric direct-cranking starter showing its construction. (Courtesy Eclipse-Pioneer Division, Bendix Aviation Corporation)

flight. The starter should be checked for security of mounting, tightness of housing bolts, and proper safetying. If equipped with hand-cranking extensions, the brackets and supports should be examined for security of mounting and general condition. A check should always be made for the presence of the hand crank. At the 50-hr. period, a careful check should be made to determine the condition of the brushes to see that they are bearing properly on the commutator. Worn or binding brushes should be properly adjusted or replaced. The general condition of the brush holders and commutator should be checked. The hand-crank systems should be lubricated, and the condition of the solenoid switches should be checked for proper operation, security of mounting, and condition of leads.

When adjusting starter solenoids, the starter jaw should be completely retracted against the baffle plate. Defective control switches or solenoids should be replaced. If the commutator is rough, out-of-round, badly scored, or burned, the motor should be replaced. Commutators may be cleaned with No. 0000 sandpaper. Never use emery cloth. After cleaning, the commutator should be thoroughly blown out with clean air. Some starters should have the commutators conditioned with a

Fig. 45. Armature and commutator for an engine-driven generator. Commutator being checked for roundness. (Courtesy Jack & Heintz., Incorporated)

special kit as recommended by the manufacturer. Brush holders and binding brushes may be cleaned with a cloth moistened in unleaded gasoline. New brushes should be given a proper seat with No. 0000 sandpaper.

Other Electrical Equipment. Other electrical equipment, such as alternators and inverters, should be checked daily or before flight to ensure their satisfactory operation. Alternators or dynamotors should be checked at the 50-hr. period for loose connections, condition of commutators, spring tensions of brushes, worn or binding brushes, security of mountings, condition of contact points, and general cleanliness. All wiring should be checked for condition of insulation and attachments. Inverters and auxiliary motors and generators do not usually need reconditioning except at regular overhauling periods. They should, however, be given a visual check at the regular inspection periods to determine their general condition, proper functioning, and operation. They should be checked carefully at approximately 250 hr. Voltage-booster dynamotors may usually be operated approximately 1000 hr. before overhauling.

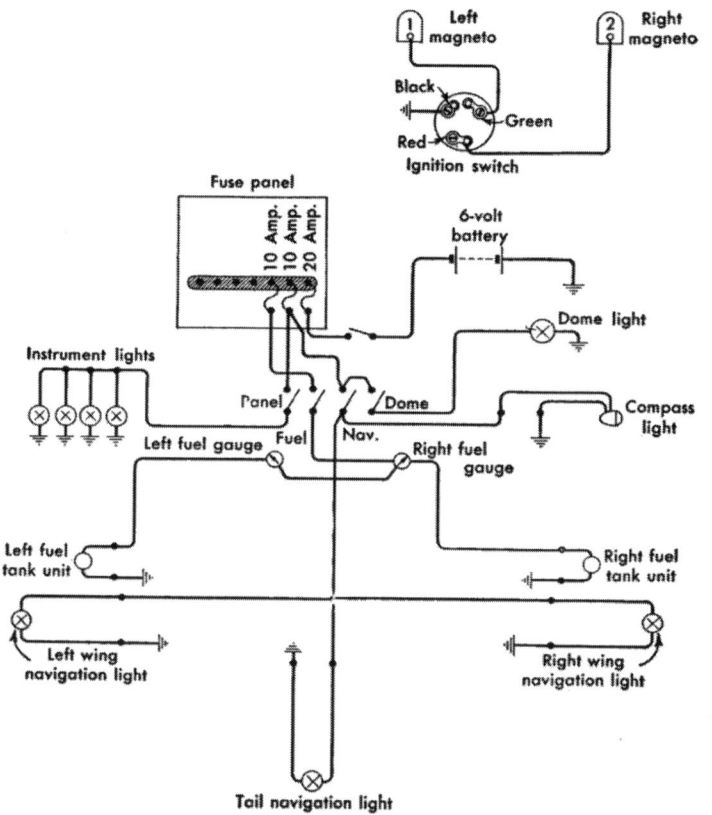

Fig. 46. A six-volt electric system. (Courtesy Consolidated Vultee Aircraft Corporation, Stinson Division)

Storage Batteries. Approximately once a week, storage batteries should be inspected for leakage of acid, and the condition of the case and sealing compound. The specific gravity of each cell should be tested, and distilled water added as required. Battery terminals should be cleaned of corrosion and the condition of leads and terminals checked.

Lights and Wiring. Daily or before flight, the proper functioning and condition of all lights, lamp rheostats, and switches should be checked to determine their proper functioning. At the 50-hr. period, all wiring should be checked for proper fastening, and the condition of bonding, insulation, and terminals. All lights should be inspected as

Fig. 47. An aircraft storage battery. (Courtesy Delco-Remy Division, General Motors Corporation)

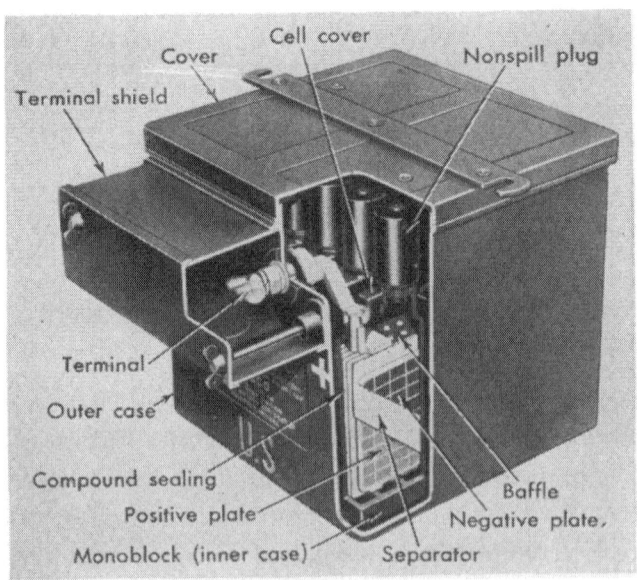

Fig. 48. A cutaway view showing the construction of an aircraft storage battery. (Courtesy Delco-Remy Division, General Motors Corporation)

to security of sockets and proper fastening, and switches should be examined for proper operation, cleanliness of contacts, and security of mounting.

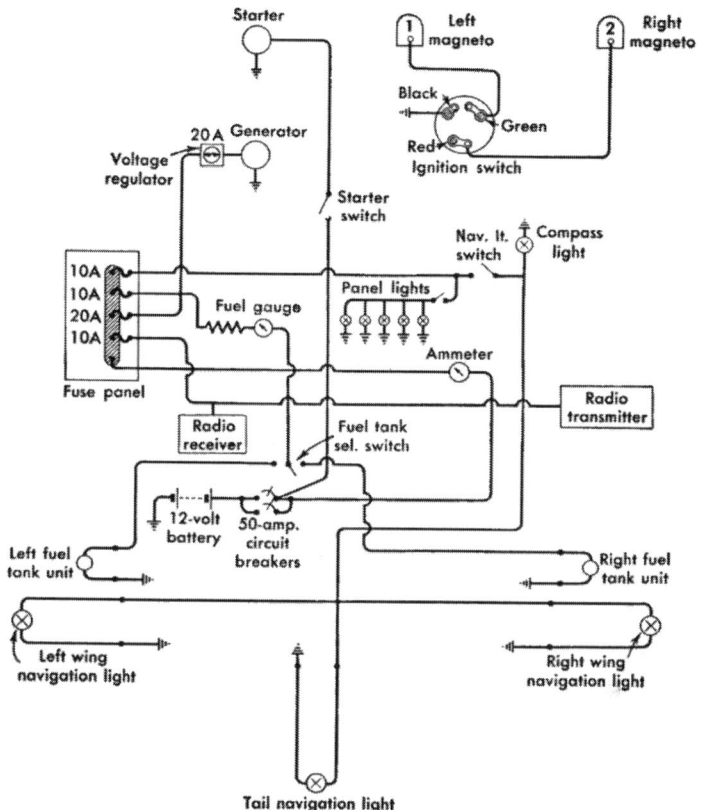

Fig. 49. A twelve-volt electrical system. (Courtesy Consolidated Vultee Aircraft Corporation, Stinson Division)

Oil Systems, Fuel Systems, and Induction Systems. Daily or before flight, the entire oil system or lubrication system should be given a visual inspection for any indication of oil leakage, looseness of oil lines and connections, tightness of oil plugs, and proper safety. The manual control handle of the Cuno oil filter should be given one revolution in a counterclockwise direction at intervals of not more than 10 hr.

If the system is equipped with an oil-dilution arrangement, its operation should be checked. The proper operation of oil-pressure indicators, oil-level gauges, and oil-pressure warning switches should be checked if these are part of the system installed. Oil coolers and shutters should be examined for cleanliness, damage, leaks, proper operation.

and security of mountings. Oil coolers should be carefully checked, and the oil-filler cap properly secured.

At the 25-hr. period, all oil lines should be checked for dents, leaks, wear, or other damage. Particular attention should be paid to the connections for any indication of failure from vibration or fatigue. The oil filter should be removed and cleaned. At the 50-hr. period, the oil-pressure warning switches should be carefully examined for leaks, tightness of connection, and condition of electric wiring. The entire system should be drained and thoroughly flushed out with the proper flushing compound, and the oil tanks should be inspected for security, condition of padding, condition of supporting straps, leakage, and signs of cracks. All lines should be checked to see that they are properly anchored. The control linkage and valves of the oil dilution control should be checked for any looseness or binding. The Cuno filters should be cleaned after the ground run-up prior to flight and after the first five hours of flight operation.

Fuel Systems. Daily or before flight, all accessible fuel drains, strainers, and sumps which have drain cocks should be drained. After draining, all drain cocks and plugs should be properly tightened and safetied. The fuel cocks should be turned on and, if the airplane is equipped with a pressure system, the fuel pressure should be tested by means of the hand pump. All fuel cocks should be carefully checked for signs of improper operation, particularly for signs of binding. If included in the system, fuel-pressure signals should be examined to see that they are operating properly.

The fuel level in all tanks should be carefully checked. Filler caps should be installed and properly secured. At the 25-hr. period, the fuel drain and overflow lines should be inspected for insecurity, breaks, or stoppages. All removable strainers should be removed and cleaned. The fuel pump should be checked for security of mounting and proper safetying. All vent openings in relief valves should be cleaned, using a proper-sized drill or plug gauge. If equipped with a pressure system, the fuel pressure should be built up with a hand pump and the entire system checked for leaks, cracks, security of lines and fittings, wear at mounting points, and to see that the warning-signal electrical connections are in good condition.

At the 50-hr. period, the fuel-pump flexible drives should be lubricated and the entire fuel system examined for general condition. The fuel-cock controls should be checked for backlash and drag. The taper

70

pin in the fuel-cock control linkage should be checked for looseness and proper safetying. After about 200 hr. of service, the fuel-pump and transfer-pump coupling should be carefully examined for wear and the shaft should be properly lubricated.

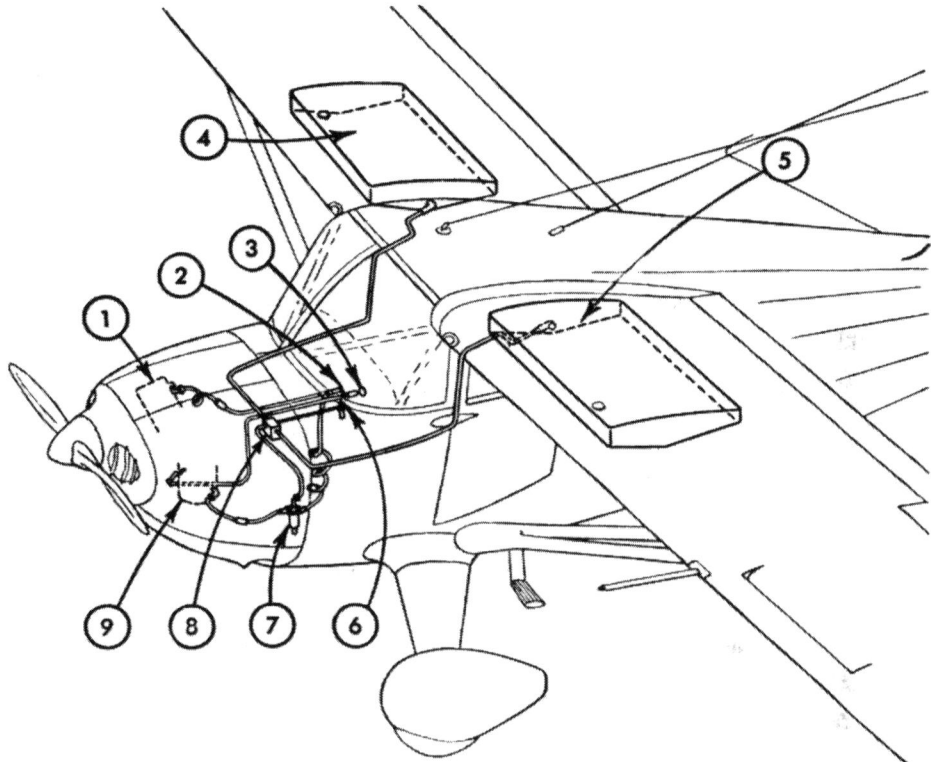

Fig. 50. A typical light-airplane fuel system. (1) Cylinder right rear; (2) primer control; (3) throttle control; (4) fuel tank—right side; (5) fuel tank—left side; (6) control handle for shutoff valve or selector valve; (7) fuel strainer; (8) selector valve (single-tank installation); (9) carburetor. (Courtesy Consolidated Vultee Aircraft Corporation)

Fuel Tanks. Daily or before flight, accessible drains and main strainers should be drained and resafetied. A visual inspection of that part of the fuel system open to such examination should be given. At the 50-hr. period, all fuel tanks, their fittings, and fuel-duct connections should be given a thorough inspection, also the anchorage of the tanks, padding, supporting straps, and gaskets; and all fuel tanks should be drained and refilled and checked for proper capacity as indicated on the fuel-level instruments.

Carburetors. Daily or before flight, the proper operation of the throttle and mixture-control rods should be examined before starting

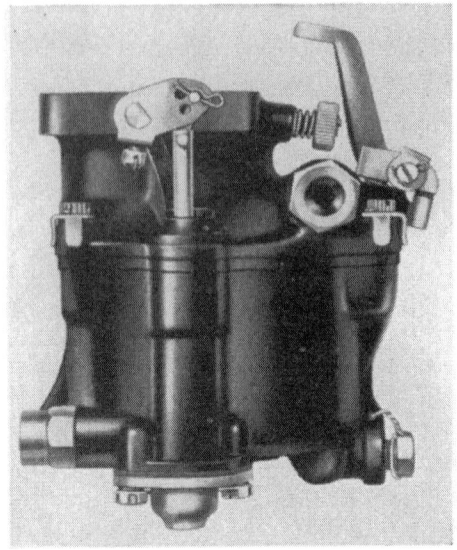

RIGHT   LEFT

Fig. 51. Right and left side view of a typical carburetor such as is used on fifty to seventy-five-horsepower aircraft engines. (Courtesy Marvel-Schebler Carburetor Division, Borg-Warner Corporation)

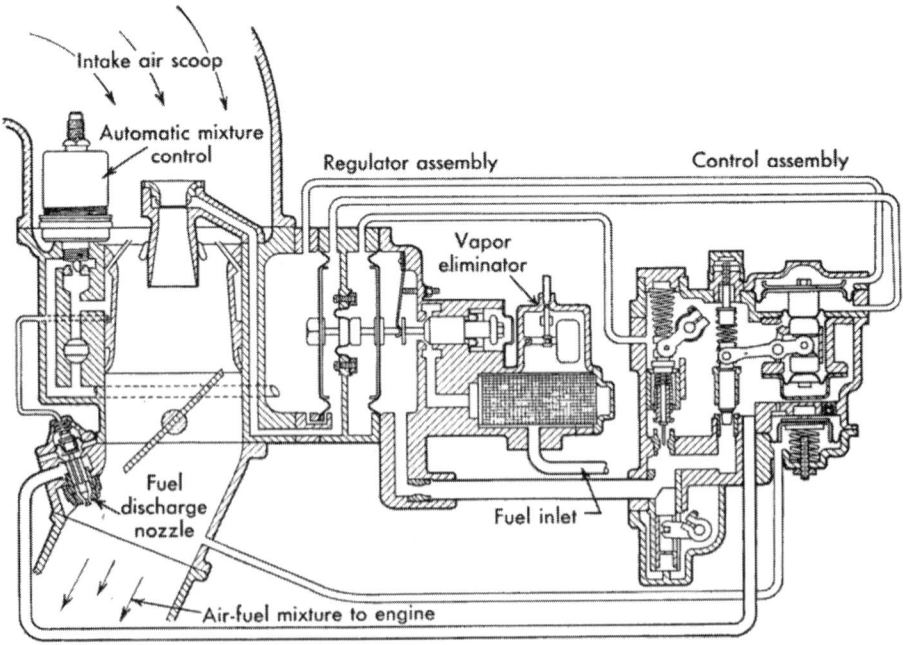

Fig. 52. An "injector" or "injection" type of carburetor.

the engine. If equipped with a pressure system, the wobble pump and engine fuel pump should be inspected. The fuel-pressure gauge and carburetor air heater should be checked. All safety wiring on the carburetor and the carburetor fuel-line connections should be checked for fuel leakage. The throttle and mixture-control connections should be checked for slack or tightness and proper safetying.

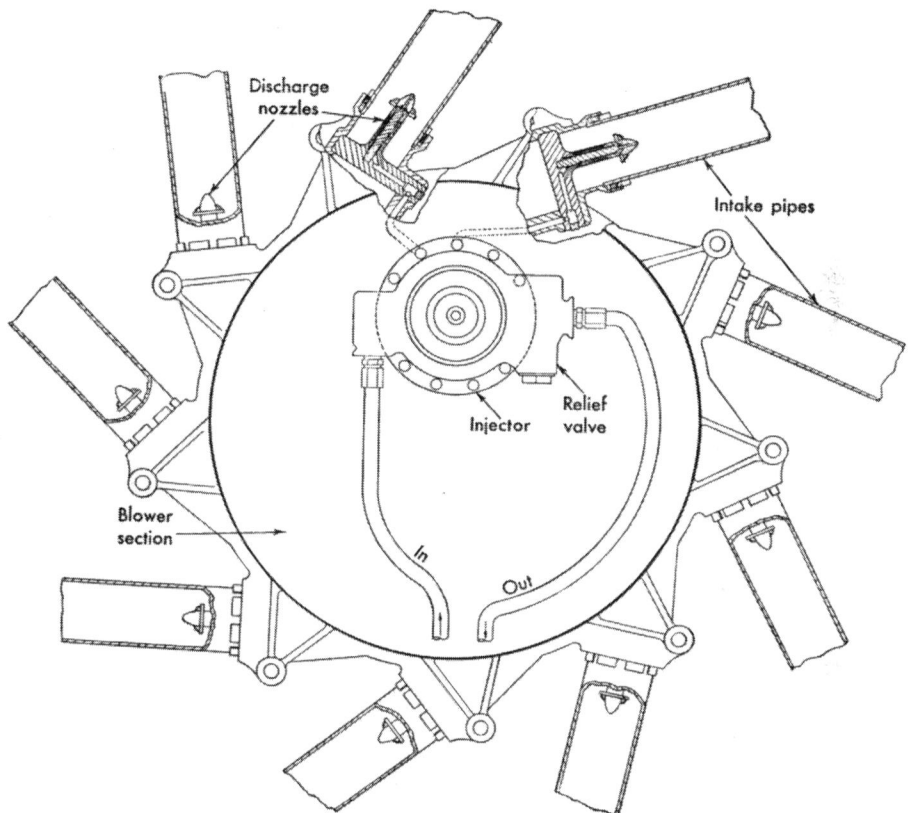

Fig. 53. A schematic drawing showing a fuel-injector system, rear view.

At the 25-hr. period, the carburetor parting surface should be inspected for leakage. The float and chamber strainers should be flushed, and the strainers cleaned. The throttle-control-bearings, exposed economizer and accelerating-pump operating parts should be lubricated. All nuts and bolts on the carburetor and on the air scoop should be tight, and all safety wiring should be carefully checked.

If the engine is equipped with a fuel-injection system, all parts of this system should be examined daily or before flight for any indication of leakage. The injection system should be examined for security of mountings, safetying, and any indication of failure. Fuel screens should be cleaned at regular intervals, and parts equipped with drain plugs should be drained daily or before flight. The plugs should be carefully replaced, tightened, and safetied.

Manifold, Turbo-Superchargers, and Regulators. Daily or before flight, the supercharger exhaust, air-induction intercooler, and control

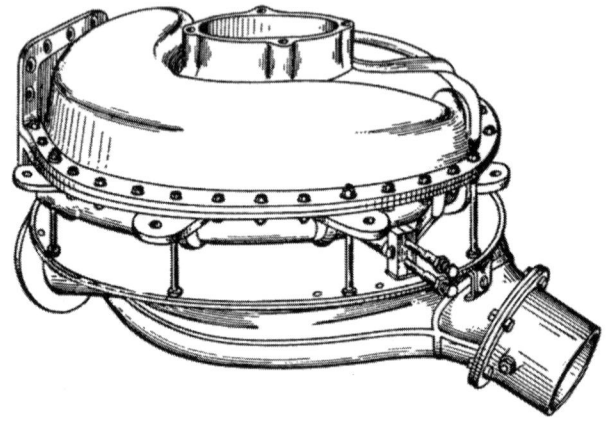

Fig. 54. A turbo-supercharger. (Courtesy Minneapolis-Honeywell Regulator Company)

system should be carefully checked for mounting and for any evidence of failure. The turbine wheel should be inspected for proper clearance, end play or side play, or other defects. The turbine wheel should be carefully inspected for cracks or any indication of failure. The exhaust waste-gate system should be checked for free movement of controls. Penetrating oil should be applied to the waste-gate moving parts.

The carburetor heater and air scoops should be checked for security of attachment. The manifold should be checked for blown gaskets and security of fastenings. The turbine wheels should be examined for freedom of operation. The operation of the entire supercharger system should be checked during the engine run-up. At the 50-hr. period, the entire exhaust system should be inspected for blown or leaky gaskets, loose or broken studs, and any other damage to the manifolds. The supercharger should be lubricated and checked for clearance, and the air-pressure lines from the engine induction system to the regulator should be examined for leaks. The pressure regulator should be examined for security of mounting, also for attachment of the oil-pressure and drain lines. The entire supercharger system should be checked for proper operation and any indication of failure. The superchargers usually operate approximately 500 hr. before needing overhauling.

V PERIODIC INSPECTIONS

The Civil Aeronautics Administration requires that certain regular inspections, commonly called periodic inspections, be made of all certificated aircraft and their component parts. These inspection periods are the "daily" or "before-flight inspection," the "25-hr. inspection," and the "100-hour inspection." Usually at the 50-hr. period, a more thorough 25-hr. inspection is made. This is sometimes called the "50-hr. inspection." The daily or preflight inspection includes certain specified items to be inspected. This inspection is entirely visual. The 25-hr. inspection is largely visual and includes items which may be inspected without disassembly. This inspection, however, may include items which require the removal of the cowling and inspection plates or the opening of inspection apertures.

The 100-hr. inspection must be made by a certificated mechanic and includes many items which require the removing of inspection plates, fairings, cowls, and accessories. The airman making any inspection does, however, not confine the inspection to the specific items named for inspection. It should become the habit of the person making inspections to examine all parts of the airplane at any convenient time for indications of failure, looseness, damage, or excessive wear.

There are certain forms prescribed by the Civil Aeronautics Administration which must be filled out by the person making an inspection. Certain of these forms become a part of the engine or aircraft logbook, others are forwarded to a representative of the Civil Aeronautics Administration.

Service, Inspection, and Related Maintenance. This section covers the work ordinarily performed by the operating organization. In this section are included such items as periodic inspection, cleaning, service, adjusting, lubrication, and such maintenance work as is done in connection with routine maintenance and service for the operation of the aircraft.

Preflight Inspection and Maintenance.

1. A preflight check should always be made of the quantity and quality of the fuel and oil in the tanks.

2. All cowling should be inspected for proper fastening; all fairing should be inspected as to its fastening and proper installation; all doors, covers, and inspection holes should be inspected to see that they are properly fastened and secured.

3. Before starting an engine, the propeller should be pulled through several times by hand to clear the cylinders of any accumulated oil or fuel. It may be necessary to remove the lower intake pipes or spark plugs to assure proper drainage.

4. During engine warm-up, the engine instruments should be carefully checked for proper functioning and readings.

5. After the engine becomes warm but not hot and while operating at slightly above idling speed, the ignition switch should be turned to the OFF position. If the engine does not stop at once, it should be stopped by turning off the fuel supply and the ground connections from the magnetos should be checked.

6. The fuel system, including the primer, should be carefully checked for leakage. Care should always be taken in touching or moving a propeller, particularly on a hot engine, to be sure that the propeller is entirely clear, as the engine may fire. This is particularly true if there is any defect in the ground connections of the magneto.

After-Flight Inspection and Service. Immediately after an aircraft has completed a flight, the following should be accomplished as soon as possible.

1. All oil and fuel tanks should be inspected and filled to the proper level.

2. Enough cowling should be removed to check carefully for any fuel or oil leaks. The wiring should be gone over for signs of failure or looseness. The fuel lines, connections, and attachments of the engine accessories and manifolds should be inspected.

3. When the airplane is through flying for the day, the propeller should be cleaned and inspected for nicks, cracks, or looseness on the hub. The metal parts of the propeller should be lightly coated with clean lubricating oil by wiping with a clean oily rag.

Daily Inspection. At least once a day, preferably before the first flight of the day, the following should be performed.

1. Remove the engine cowling.

2. Inspect the engine mount for cracks or signs of failure, and check the engine brackets and mounting pads for general condition and security of attachment.

3. Check for damaged or broken cooling fins and be sure that the cowling is not rubbing on any part of the engine and that the cowling pads are in good condition.

4. Check the manifolds and exhaust pipes for general conditions. Inspect the manifold gaskets and look for missing or loose bolts, nuts, or broken lugs.

5. Check for any indication of oil leakage.

6. Inspect the engine controls, assemblies, and linkages for proper functioning, safetying, and general condition.

7. Check the engine instruments for proper operation. Be sure that all the pointers register zero when the engine is not running and that they cover the proper operating range with the engine in operation. Check for broken or loose glass covers, and general condition and proper installation of instruments.

8. Inspect the oil and fuel lines, being sure that the vents are open. Inspect the fastening of the overflow lines, fuel lines, and oil lines, and be sure that all lines are properly secured and that there is no indication of failure, kinking, or rubbing.

9. Drain the fuel strainers and resafety.

10. Check the carburetor for tightness of mounting, leakage, binding, leaking connections, and proper safetying.

11. Inspect the carburetor air screens and clean them, if necessary.

12. Check the oil and fuel drain plugs for tightness and proper safetying.

13. Check the magnetos for cracked housing, looseness of supports, safetying, and tightness of all wiring connections.

14. Check the spark-plug wiring for damage or looseness, and check the spark plugs for cracked porcelain, condition of shielding, and any signs of leakage around the plug.

15. Replace and properly fasten the engine cowling. At the end of 10 hr. of service, the Cuno oil filter handle should be rotated. This handle should be rotated at least once during the first hour of engine operation after the engine has been overhauled or the oil changed.

Twenty-Five Hour Inspection and Service. After 25 hr. of service, the engine should have the following inspection and service as required.

1. The spark plugs should be thoroughly inspected and cleaned, and the points checked for adjustment. Where needed, the proper adjustment should be made. All spark-plug terminal connections should be checked for security, safetying, and signs of wear. The spark plugs are not usually removed for the sole purpose of checking the gaps. The gaps should be adjusted in accordance with the manufacturer's specifications.

2. All fuel lines and the carburetor should be inspected for leakage, rubbing, or signs of failure. Drain plugs, passage plugs, and parting surfaces between body castings should be carefully examined for condition and tightness. The throttle and mixture control should be examined carefully throughout to determine the operating condition, tightness, and safetying. Inspect all safety wiring on the fuel system.

3. The oil should be changed. The sump oil screens should be removed and cleaned, and the sump chamber thoroughly flushed out and cleaned. If metal particles are found in the screens, the engine should not be operated until their source is found and necessary corrective measures taken. The draining and cleaning of an oil system should be done while the engine is still hot. All oil lines should be inspected for security of fastening, chafing, vibration, dents, or any signs of failure. Hose connections and clamps should be inspected for general condition.

Cuno Oil Filter Inspection and Service. The Cuno oil filter should be carefully examined at each 25 hr. inspection period.

1. All cartridges should rotate through 360° with a maximum torque variation of 50 per cent. Hard spots or points of catching are reasons for rejection.

2. All cleaner blades must be straight and flat. The blades should not show angular displacement in excess of approximately 8° from the mid-position when the cartridge is rotated. Bent or torn blades are cause for rejection. In an emergency, a torn or badly bent blade may be carefully removed. When this is done, it must be assured that all broken parts are removed from the filter and that no other part is damaged. This filter should be used only until a replacement can be made.

3. All discs must be flat and evenly spaced. The discs should be free from all burrs or nicks.

4. The Cuno filter should be cleaned every time it is removed from

an engine. This cleaning is done by washing in kerosene, gasoline, a 50 per cent mixture of carbon tetrachloride and benzol, or another similar noncorrosive solvent. All foreign matter should be carefully removed from the filter. Rotating the filter while in the solvent will assist in cleaning. Hard-edged or pointed tools should never be used to scrape or pick at the cartridge. As soon as the cartridge is cleaned, it should always be dipped in clean engine oil.

5. If the cartridge cannot be rotated through 360° and inspection reveals no mechanical defect, the packing gland may be loosened or the filter placed in a solvent and rotated gently, but firmly. This permits the solvent to work in and free parts which may be stuck with sludge. If this treatment fails to free the cartridge, the unit should be overhauled or replaced.

Valve Mechanism Inspection and Service. A complete inspection of the valve-operating mechanism should be made every 25 hr. of engine operation. This inspection should be made any time that rough running of the engine is encountered, if the cause cannot be traced to any other source. This inspection of the valves should include the removal of the rocker-arm cover; checking for proper clearance; adjustment; checking of the valve springs and their fastenings; and the proper lubrication of rocker arms, valve stems, and push rods.

Manifold Inspection and Service. The exhaust stacks and intake pipes should be inspected for cracks, chafing, and general condition. The intake-pipe gaskets and packing should be carefully examined for any evidence of leakage. The flanges should be tested for tightness, and the packing should be replaced if any leaking is suspected.

Propeller Inspection and Service. The crankshaft thrust-bearing nut should be checked for tightness. The propeller should be checked for tightness on the hub and for proper alignment. If any looseness is detected, the nut should be tightened in accordance with the manufacturer's recommendation.

Fifty-Hour Inspection and Service. At the end of 50 hr. of operation, the following should be done in addition to the regular 25-hr. inspection and service.

The ignition and electrical systems should be inspected and serviced as follows.

1. Remove the breaker from the magnetos and clean the breaker housing.

2. Inspect the magnetos for

    a. Damaged cam followers
    b. Damaged breaker felts or cushions
    c. Weak or broken breaker-arm springs
    d. Worn or loose cams
    e. Excessive lubrication, but lubricate if necessary.

All oil should be cleaned from the contact points. Be sure the entire breaker mechanism is clean and securely mounted and safetied.

3. Inspect the magneto distributor for

    a. Cracked or damaged rotor
    b. Sticking or broken brush
    c. Signs of arcing or shorting
    d. Mounting screws for tightness
    e. Worn or damaged brush rotor or other parts
    f. Breaker points and check points for adjustments in accordance with the manufacturer's recommendation.

When adjusting the contact point of the pivotless breaker, the adjustment must be made so that the contact points open when the chamfered tooth of the large distributor gear is opposite the timing pointer on the inside of the main cover. The points on this type of breaker are not adjusted for any fixed clearance when open. To check the adjustment, turn the crankshaft until the chamfered tooth of the large distributor gear is opposite the timing pointer on the inside of the main cover. When in this position, the contact points should just begin to open. A service tolerance of $\frac{1}{16}$ in. is allowed. That is, the timing marks may be off as much as $\frac{1}{16}$ in. when the contact points open before adjustment is necessary. When inspecting the contact points for any reason, do not raise the breaker main spring beyond a point giving $\frac{1}{16}$ in. clearance between the contact points. Bending the spring farther than this will weaken it and cause unsatisfactory magneto performance. If the contact points are pitted or burned, a file should not be used for dressing them down. The contact points must be removed and dressed down by a special file and stone. During this dressing down, they should be supported in a suitable block. If inspection shows that adjustment is necessary, the manufacturer's recommendation should be followed.

One-Hundred-Hour Inspection and Service. In addition to the regular 25-hr. and 50-hr. periods of inspection and service, the following should be accomplished after the engine has been in service 100 hr.

PERIODIC INSPECTIONS

1. Spark plugs should be replaced with new or reconditioned plugs of an approved type.

2. The parting surfaces in the body of the carburetor should be inspected for leakage. All nuts and bolts of the carburetor should be inspected for tightness. The fuel strainer plug and strainer from the carburetor should be removed and cleaned. Water and sediment should be flushed out of the carburetor and fuel lines by allowing fuel to flow through the strainer and drain-plug openings. Remove the air intake if it interferes with the removal of the drain plug or strainer. Strainer assemblies, strainer plugs, or plugs marked "drain" should be replaced if found in bad condition.

Three-Hundred-Hour Valve Inspection and Service. A complete inspection of the valves and their operating mechanism should be made at the end of each 300 hr. of engine service. This inspection will include the following.

1. The rocker boxes and covers should be removed and thoroughly cleaned and inspected.

2. The rocker arms should be examined for freedom of movement. If any binding or restricted movement is found, the rocker arms should be removed and the bearings inspected. The rocker arms should be inspected for cracks or any interference of adjoining parts, or excessive end or side play. Rocker-arm rollers or bearing surfaces should be inspected for excessive wear. The valve rocker rollers should be inspected for any indication of binding, flat spots, and proper safetying of the roller axle.

3. The push rods should be carefully inspected for any signs of binding or poor seating of the ball ends. The hose connection on push-rod housings should be replaced if found to be in poor condition. If the push-rod housings have any other type of connection, these connections should be inspected for tightness and any evidence of leakage.

4. The valve springs should be inspected with the valves closed. The valve springs should be tested for proper pressure and proper seating, and the retainer washers or other fastenings should be checked to see that they are in place.

The safetying of the valve collar lock should be carefully examined. Carefully inspect springs for any signs of breakage or rusty spots.

5. The valves and valve-stem bushings should be examined for signs of excessive wear or improper seating.

6. Inspect the valve-rocker shaft to determine that it is properly

locked in position. If there are signs of oil leakage around the shaft, it should be removed and the oil seals should be replaced. If the shaft has been removed for any reason, the oil should always be replaced.

7. The valve clearance should be checked and, if necessary, the proper adjustment should be made.

8. All parts of the valve-operating mechanism should be checked to see that they are properly lubricated. New gaskets should be used whenever the rocker-arm box covers are removed.

VI TROUBLE SHOOTING

Trouble Shooting and Service. Aircraft-engine failures or troubles may be held to an absolute minimum by careful inspection, maintenance, and service. It is essential that every member of the crew responsible for these operations be thoroughly familiar with all parts of the power plant and the conditions under which they operate. Power-plant failures may be divided into structural failures and operation failures.

Structural failures are caused by the actual breakage of some part of the structure. This breakage may be caused by unusual vibration conditions which set up fatigue in various metal parts, or the vibration may be severe enough in itself to cause overstressing and breaking of parts. Mechanical failures or structural failures are few and far between. Most forced landings due to power-plant failure can be directly traced to faulty maintenance or service.

Structural failures are usually caused by operating the engine above its rated r.p.m., operating under extreme temperature conditions, using a low-grade fuel which causes overheating and detonation, or continued abuse without proper maintenance.

Operation failures may be usually traced to faulty service or maintenance. Faulty service or maintenance may lead to the partial or total failure of the ignition system, the lubrication system, or the fuel system. The lack of oil or fuel, or the use of an improper grade of oil or fuel, has been the cause of many forced landings or partial failure of engines in service. Lack of careful inspection and service is one of the main causes for ignition failure.

While structural failures usually occur suddenly, operational failures almost always give some indication by which the pilot who is thoroughly familiar with the operating conditions of the engine may be forewarned, allowing a comparatively safe emergency landing. Operational troubles may be indicated by such warnings as loss of power, abnormal temperatures, irregular operation of the engine, rough operation of the engine, and vibration. Indications of both structural and operational failures should always be detected by the skilled mechanic in charge of maintenance and service.

Indications of ignition trouble may be shown by a difference in the operation of the engine with the ignition switch in its different positions. Instruments are largely installed to give warning of improper functioning of the various parts of the aircraft.

Service personnel, as well as operating personnel, who are not thoroughly familiar with all parts of the structure and its operations have a tendency to decide that the trouble exists in one place and will proceed on that assumption rather than check all possibilities. For example, uneven firing of the cylinders may be caused by the improper functioning of the ignition system, carburetor, fuel system, induction system, or the lack of compression in one or more cylinders due to warped or sticking valves or worn rings.

The fault may also be in the fuel itself. Even though it was decided that the ignition system was at fault and upon careful inspection some part of the ignition system was found improperly functioning, it would still be necessary to check all possibilities before the engine is returned to service. In making the inspection of a power plant, a systematic procedure should be followed. It has been recommended that the following order be used in making the service inspection:

1. Check the engine for proper compression.
2. Check the lubricating system.
3. Determine the proper functioning of the cooling system.
4. Thoroughly check the carburetor, fuel, and induction system.
5. Carefully inspect the ignition system and timing.

The engine should be carefully checked while in operation for any signs of improper operation or functioning. To assist the mechanic in his work the pilot should note on his report any indications which appeared during flight.

The propeller should be turned over by hand several times in the direction of normal rotation to test each cylinder for its proper compression. If the missing occurs in any one cylinder, the ignition system, carburetion system, lubrication system, cooling system, and timing can usually be eliminated because the trouble is probably within that cylinder only.

Certain well-known signs should be familiar to the mechanic whose responsibility it is to check or inspect an engine. For example, leaking valves produce a hissing or blowing sound as the engine is brought up to compression in that particular cylinder. Valves that are stuck open will allow the piston to pass over the compression point freely. Leaks in the induction system can usually be determined by hissing noises. A leaky intake valve will usually make a hissing noise which will be heard in the carburetor or scoop, or in the vicinity of the carburetor. A leaking exhaust valve causes a hissing noise which may be heard in the exhaust manifold or at the end of the exhaust stack. A hissing heard in the crankcase may indicate defective rings, excessive blow-by caused by a distorted cylinder, or a cracked piston. Of course, the compression in the cylinder may be measured by a compression gauge. When a compression gauge is used, the engine should be rotated to normal cranking speed with the throttle wide open.

When loss of compression is caused by leakage, the removal of the rocker-arm box covers and a check of the valve operation may assist in locating the trouble.

Engine Starting. Trouble is often encountered in starting an engine but, if there is plenty of fuel and all parts are properly adjusted, the modern engine should start promptly. If the engine is equipped with a mixture control, it should be set in the full-rich position. The air-heater control should be set in the full-cold position. If the engine is equipped with a two-position, or constant-speed, controllable-pitch propeller, it should be started with the propeller in low-r.p.m. (high-pitch) position to prevent starving the engine of oil. After operating the engine for about 30 sec., the propeller control should be moved to the high-r.p.m. (low-pitch) position. The propeller should be in low-pitch position during the warm-up period.

If the engine is warm, the throttle should be cracked (opened slightly). The starter button should be pressed if the engine is equipped with a starter. As the engine starts to revolve, the ignition switch is placed in the ON position. The same general procedure should be used if the engine is pulled through by hand.

It is usually good practice to pull the engine through about two complete revolutions with the throttle cracked before turning on the ignition switch. When starting by hand, the propeller should be set in the position to pull through, and the person pulling the propeller should then stand clear while the switch is turned on. Care should be taken not to "pump" the propeller after the switch is turned on. The propeller should be pulled through as rapidly as possible. Great care should be taken to be clear of the propeller when the engine starts.

If the engine is cool, but not cold, the primer should be pumped three or four strokes while turning the engine over by means of the starter or by pulling through by hand before the switch is turned on. If the engine is cold, the throttle should be left tightly closed with the switch in the OFF position. The engine should be turned over three or four complete revolutions either by use of the starter or by hand. While the engine is being turned over, the primer should be operated for four to six full steady strokes. The primer should be left in the full-out position and the throttle cracked. Close the primer as soon as the engine starts.

When the starter is used, turn the switch to the ON position (to BOTH), and pump the primer as necessary until the engine runs smoothly. The primer injects gasoline directly to the intake ports. If the engine is pulled through by hand, the use of the primer assists in furnishing enough fuel when the engine starts to keep it running. The primer should be closed and locked as soon as its use is no longer needed. If the primer is not closed and locked in a closed position, the engine may draw gasoline through the priming system, increasing the fuel consumption and interfering with satisfactory operation. Unless the manufacturer's instructions so indicate, or the engine is not equipped with a primer, the throttle should never be pumped.

Some engines are equipped with both a magneto and a battery ignition system. On engines so equipped, the engine is usually started on the battery system. When ready to start, the switch is placed in the BAT position. Some engines, although equipped with two magnetos, have an arrangement whereby the engine may be started on an auxiliary battery ignition system.

If the engine fails to start, do not continue to grind away on the starter or use the primer excessively. Repeated or long attempts to start an engine without checking to determine the reason for the failure may cause the lubricating oil to be washed off the piston and cylinder walls. If this happens, the pistons and cylinder walls may be scored before the oil in the crankcase is warm enough to lubricate them properly. If the engine fails to start, the cylinders may have become flooded. If this cause is suspected, the engine should be rotated backward through six or eight complete revolutions with the throttle wide open. If a starter is used, the engine may be rotated several times in the normal direction of rotation by the use of the starter with the throttle wide open and the ignition switch in the OFF position.

The oil pressure should show on the oil-pressure gauge within 30 sec. after the engine has been started. If the pressure is not indicated, the engine should be stopped and an investigation made to determine the cause.

The engine should be run for a short period at the lowest practical idling speed. The speed should then be gradually increased to 800 or 900 r.p.m. and operated at this speed until the oil has reached the proper operating temperature. The engine should not be operated at full throttle on the ground except to determine whether or not it is turning up to the full-rated r.p.m. and to check the operation of the engine instruments. Under no circumstances should the engine be operated at full throttle on the ground for more than a few seconds at a time.

The ignition switch should be checked in each position, and the greatest loss in r.p.m. should not exceed approximately 100 when operated on either magneto alone or on a battery system. The engine should be checked on all fuel tanks to be sure that all fuel lines are open and free of water. The carburetor mixture control should be in the full-rich position for take-off.

When the normal starting operations have been performed and the engine fails to start, a careful check should be made to determine the cause. If the weather is damp or cold, the front spark plugs should be removed and slightly heated with a torch. This should be done whenever atmospheric conditions are such that condensation on the plugs might occur. Condensation of moisture on the plugs is one of the most common causes for failures to start. The spark plugs should also be checked for improper gap or fouling by oil. Check to make sure that the fuel cocks are operating properly and a plentiful supply of fuel is being furnished to the carburetor.

Check the gas flow at the carburetor, also, and examine the fuel strainer. To do this, the strainer should be removed and both sides of the screen carefully examined. The gasoline lines should be checked for damage that might restrict the flow of fuel. The fuel-tank vents should be examined to be sure that they are not clogged. If the flow of gas to the carburetor is plentiful, the carburetor itself should be checked. The carburetor should be examined for clogged screen, clogged jets, and sticking float or needle valves. In cold weather, water may collect in the carburetor or fuel lines and may become frozen.

If starting on a battery ignition system, the battery may be low. Make sure that it is fully charged by checking with a hydrometer. All wiring and terminals should be carefully examined. If the ammeter shows a heavy discharge upon turning on the switch, a short in the wiring or ignition system is indicated. If no discharge is shown on the ammeter, the engine may have stopped with the points open. If, however, the ammeter fails to show any discharge or movement of the needle when the engine is rotated, an open circuit is indicated. Dirty points may also be indicated.

The plugs should be checked to see that a spark occurs. If a spark cannot be obtained at the plugs, trace the current from the battery or magneto to the distributor by the use of a voltmeter. It may be necessary to use an auxiliary source of current, such as a battery, to trace circuits. To do this, after the proper connections are made, proceed as follows for most normal installations:

1. Check to see that there is current to the negative side of the ammeter on the instrument panel.

2. Check the positive side of the ammeter for current.

3. Check the current from the switch to the coil and from the coil to the distributor.

4. The breaker points, condenser, and ground connections should then be checked.

5. If the primary circuit is found to be functioning properly, check the secondary circuit.

6. Check the ground connection of the secondary coil.

7. Be sure an electric current will flow through the high-tension coil.

8. Check the high-tension lead to the distributor.

9. Check the leads from the distributor to the spark plugs.

10. Examine *all* wires to see that they are in good condition and are securely fastened to the terminals.

In cold weather, the oil may be congealed in the engine, making it impossible for the starter to rotate the engine at the proper speed. Turning the propeller over by hand through several revolutions before attempting to start the engine will assist in relieving the starter of this additional load. If the weather is extremely cold, it may be necessary to preheat the oil or even heat the engine itself.

The carburetor-mounting-flange, intake-pipe connections, intake-pipe packing, intake manifolds, and priming hole plug should all be checked for air leaks. Inspect for stuck valves or improper valve clearances.

Common Causes of Failure of Engine. It is important that every maintenance and service mechanic be an expert trouble shooter. Trouble shooting is a common term used to indicate the method by which the reason for faulty operation of the engine or its accessories is located. Listed below are the common causes for failure of the power plant and its accessories and the common causes for such faulty operation. It is important that, when any type of faulty operation is found, the mechanic check all of the probable causes. There is always the chance that the fault may be caused by a combination of the probable causes given.

Low Oil Pressure. Low oil pressure may be caused by any of the following conditions.

1. Dirt in the oil strainer. (The screen should be removed and thoroughly cleaned.)

2. The oil temperature may be too low. (Check the operation of the oil-temperature gauge.)

3. Poor connections causing leaks in the oil suction line. (All lines and connections should be carefully checked.)

4. Low oil supply. (The level of the oil in the oil tank should be determined.)

5. The ball in the pressure-relief valve may not be properly seating. (There may be dirt or particles of carbon between the ball and the seat, allowing oil to escape.)

6. A defective oil gauge. (Check with a new gauge.)

7. Plugged, kinked, or restricted oil lines. (The oil lines should be carefully examined for these defects.)

8. The pressure-gauge line leading to the oil-pressure gauge may be broken, kinked, or clogged. (Be sure this line has no sharp kinks. Remove and clean, if necessary, and fill with light oil.)

9. Oil may be of too low viscosity. (Check to see that the oil is of the proper viscosity and grade.)

10. The pump itself may be worn or defective. (Check the pump to see that it is delivering the proper quantity of oil at the proper pressure.)

11. The pressure-relief-valve spring may be broken or improperly adjusted. (Disassemble the pressure-relief valve and check all parts.)

12. The engine bearings in the pressure system may be badly worn, or a broken bearing may be allowing the oil to escape.

13. A plug may be out or loose in the pressure system. (It may be necessary to dismantle the engine to correct this fault.)

High Oil Pressure. 1. The oil may be cold or of too high viscosity. (Check the viscosity of the oil in the tank.

2. The engine oil-pressure gauge may be defective. (Check with another gauge.)

3. The oil-pressure relief-valve spring may be improperly adjusted, or the ball may be stuck on the seat.

4. Heavy oil may be congealed or accumulated in the oil-pressure-gauge line. (Remove, drain, and clean the line and fill with light oil.)

High Oil Temperature. 1. The oil supply may be low. (Be sure that the tank is filled to the proper level.)

2. The oil coolant lines or coolers may be clogged or the oil cooler shutters may not be operating properly. (The air passages in the oil coolers may be clogged with insects or dirt, or the cooling air supply to the engine may be shut off.)

3. The oil may be dirty or diluted. (Replace with clean oil of the proper grade.)

4. The oil lines or oil screens may be partly clogged. (Clean all oil screens and blow out oil lines.)

5. Badly worn or stuck rings. (It may be necessary to remove the cylinders and examine the rings and valves.)

6. A badly worn master-rod bearing or any other bearing in the oil-pressure system may allow oil to escape, causing poor lubrication of the rest of the engine.

7. Faulty scavenging due to air leaks in the line from the sump or restricted line to the tank, not allowing a free flow of oil from the scavenger pump.

High Oil Consumption.

1. Worn piston rings.

2. Oil of too low viscosity.

3. Too much clearance in the master-rod bearings.

4. The oil pressure may be too high. (Inspect the relief valve for proper adjustment.)

5. Improper grade of oil.

6. Scavenger pump not operating properly.

7. Excessive operating temperatures.

8. Oil leakage anywhere in the system.

9. Worn or defective oil seals anywhere in the system.

10. Cracks in the built-in oil lines.

Overheating.

1. Mixture too lean.

2. Improper altitude adjustment particularly at low altitudes.

3. Low oil supply.

4. Improper grade of gasoline.

5. Improper ignition timing.

6. Damaged baffles or improper functioning of cowl flaps.

7. Climbing under full power at low air speed.

8. Viscosity of oil too low.

9. Defective piston rings or cylinders causing blow-by.

10. Improper cooling of oil, due to defective oil radiators.

11. Leaks in the induction system, allowing the intake of excess air.

12. Operating supercharged engines above the proper r.p.m.

13. The manifold pressure too high.

14. Faulty ignition on one set of plugs, or some other reason causing detonation.

15. The carburetor air heat may be on or excessively warm air may be admitted to the carburetor.

16. Preignition caused by carbon accumulations or overcompression of low-grade fuel.

17. Clogged air intakes to the engine cooling system.

(NOTE: With liquid-cooled engines, the following may be additional causes for engine overheating.)

1. The coolant supply may be low.

2. The circulating pump may be functioning improperly.

3. The flow of the coolant through the radiator may be restricted.

4. There may be an obstruction in the coolant line, particularly at hose connections.

5. The thermostatic controls may not be operating properly.

In all cases of overheating or engines running too cool, the temperature indicators should be checked for accuracy.

Roughness. Roughness, vibration, or uneven functioning of the engine may be due to a number of different causes. Vibration may be caused by broken or loose supports. Faulty ignition which is a common cause of roughness may be due to dirty or poorly adjusted breaker points, defective wiring, defective coil, defective condensers, or faulty timing. The magneto distributor block should be carefully checked. The distributor cap, rotor, and battery distributor, if part of the equipment, should all be carefully examined for defects, cracks, or indications of shorting. Roughness may be caused by the engine's misfiring because of restricted fuel flow. Carefully clean the fuel strainers and look for leaks in the induction system. The manifold pipes and connections should be examined carefully. See that the jets are free from dirt and that the carburetor float is at the proper level. Be sure that the carburetor air is within the temperature limits recommended for the operation of the engine.

Other causes for roughness or vibration are:

1. Propeller out of balance, loose on the shaft, or not properly mounted.

2. A bent crankshaft, or loose front main bearing or thrust bearing.

3. Engine loose on its mount.

4. Faulty valve action due to improper adjustment of rocker-arm clearances.

5. Low engine temperature; a cold engine may run "rough."

6. The mixture may be too rich or too lean.

7. Preignition or detonation may cause rough engine operation.

The propeller should be checked for balance and track. The propeller should track from within $\frac{1}{16}$ in. to $\frac{1}{8}$ in. (Check manufacturer's recommendation.) The position of the propeller on the mounting hub should be checked. Metal propellers are usually installed in line with the crankshaft throw. Wooden propellers are installed 15° to 18° ahead of the crankshaft throw. Wooden propellers may have absorbed moisture which would throw them out of balance. Changing the position of the propeller blades in relation to the crankshaft throw may correct vibration or rough running. Wooden propellers should be checked at regular intervals for balance, track, and condition of the protective covering. Damage to various parts of the propeller may cause roughness. Be sure that the crankshaft thrust nut is properly tightened. The engine mount should be checked for looseness of bolts, cracked or broken mounting lugs, cracked or broken welds, or missing bolts. The valves should be checked for proper compression, sticking, or improper clearance.

Irregular Running. Some of the most common causes for irregular running are listed below.

1. Irregular firing of one or more spark plugs. (If this cause is sus-

pected, the plugs should be removed and cleaned, the gaps adjusted, and the plug "bomb-tested.")

2. Ignition points. (The points should be checked for pits, burning, and proper gap.)

3. Faulty ignition wires. (All ignition wires should be examined for broken or burned insulation, and all connections should be cleaned and readjusted.)

4. Sticky valves. (The valves should move freely in the guides and the rocker boxes should be well lubricated.)

5. Improper grade of fuel. (Check the grade of fuel being used and be sure that it is free from water or other foreign matter.)

6. Ice in the carburetor. (Check the carburetor heat and use the carburetor heat adjustment as directed by the manufacturer's operating instructions.)

Low Power. Low power may be caused by a number of conditions of which the following are the more common.

1. Faulty functioning of the ignition system. (The drop in r.p.m. when going from the BOTH position to either battery or a single magneto should not exceed approximately 100 r.p.m. If the loss in r.p.m. exceeds this amount, the magneto or battery system should be carefully examined. Check the spark plugs, ignition wires. breaker points, coil, condenser and timing.)

2. Poor compression. (Check to see that the valves are seating properly and properly operating in the guides. Check for blow-by, faulty piston rings, or cracked pistons. Look for low compression in one or more cylinders which may be caused by: a. leaking valves; b. sticking valves, c. excessive blow-by due to faulty rings or scored or warped cylinders; d. cracked or broken pistons; e. cracked cylinder head or barrel; f. improper lubrication in the cylinder.)

3. Faulty induction. (Check for obstructions in the air induction system. Check for the full opening of throttle valve and obstructions in the screens, air cleaners, preheaters, and air scoops.)

4. High carburetor air temperature. (Check for proper indications of the air temperature gauge and proper functioning of the carburetor heat control.)

5. Ice in carburetor. (Check carburetor air temperature.)

6. Incorrect setting of propeller blades. (If the propeller blades are set at too high a pitch, the engine will not be able to turn up to its full rated r.p.m.)

7. Low grade fuel or restriction in the fuel supply lines. (Check for proper flow of fuel to the carburetor and for proper operation of the carburetor.)

Miscellaneous Causes. The following have been given as miscellaneous causes for faulty engine operation.

1. Causes of detonation
    a. Excessive cylinder temperatures
    b. Too low octane rating of fuel
    c. Mixture too lean
    d. Overspeeding engine
    e. Steep climb at low air speed
    f. Excessive blow-by (defective pistons, rings, or cylinders)
    g. Viscosity of oil too low
    h. Improper lubrication of cylinders, pistons, and rings.
2. Causes of preignition
    a. Continued detonation
    b. Excessively lean mixture
    c. Excessive cylinder temperature
    d. Spark plugs too hot
    e. Feather-edged valves or seats
    f. Excessive carbon deposits.
3. Causes of excessive vibration
    a. Propeller out of balance or not properly mounted
    b. Bent crankshaft
    c. Engine loose on mount
    d. Rocker-arm clearances not properly adjusted
    e. Engine temperatures too low
    f. Mixture too lean or too rich
    g. Detonation
    h. Preignition.
4. Causes of intermittent misfiring of engine
    a. Incorrect mixture
    b. Low-grade fuel
    c. Water in fuel system
    d. Air leaks in the induction system
    e. Erratic valve action
    f. Defective spark plugs
    g. Defective ignition harness
    h. Defective magnetos.

5. Causes of misfiring of engine at high speed
   a. Defective magneto coil, distributor, or breaker assembly
   b. Defective ignition harness
   c. Defective spark plugs
   d. Mixture too lean
   e. Engine operating temperatures too low
   f. Restricted fuel flow
   g. Clogged vents in fuel tanks
   h. Fuel-control valve not fully open
   i. Engine overheated.
6. Failure of engine to develop full power
   a. Faulty ignition units
   b. Faulty carburetion
   c. Supercharger broken or clutch slipping
   d. Improper valve or ignition timing
   e. Air leak in induction system
   f. Engine operating temperatures too low
   g. Fuel pressure too low
   h. Screen in carburetor air scoop partially clogged
   i. Throttle arm connected in such a manner that the throttle is not fully opened
   j. Improper lubrication
   k. Engine needs overhauling.
7. Failure of engine to start
   a. Ignition switch off
   b. Out of fuel
   c. Fuel-control valve not turned on
   d. Mixture control in lean or idle cutoff position
   e. Defective spark plugs
   f. Spark plugs coated with foreign matter or moisture
   g. Moisture in radio shielding
   h. Insufficient cranking speed
   i. Booster not functioning
   j. Impulse coupling not functioning
   k. Defective magnetos
   l. Incorrect throttle setting
   m. Defective primer, under- or overpriming
   n. Air leak in induction system
   o. Valves or ignition out of time.

8. Causes for complete failure of engine
    a. Fuel supply exhausted
    b. Water in the fuel system
    c. Clogged gas-tank vents
    d. Clogged fuel strainers
    e. Air or vapor lock in fuel lines
    f. Structural failure.

Engine warmup.

1. Reasons for warming up the engine before take-off
    a. To elongate cylinders and obtain correct valve timing
    b. To expand the steel and aluminum parts properly and obtain the correct clearances
    c. To warm the oil and obtain the proper operating temperature
    d. To warm the induction system and so obtain the proper distribution of the air-fuel mixture
    e. To obtain full power
    f. To check the performance of the power-plant units.
2. Reasons for the use of heavier viscosity oil in aircraft than in automotive engines
    a. Higher operating temperatures
    b. Greater clearances between parts
    c. Higher cylinder pressures
    d. Greater bearing loads.

VII ENGINE OVERHAUL

It is not possible in a text of this kind to give specific or exact directions for the maintenance and overhaul of all engines. The procedures given are general and may be applied to any engine. In the overhaul of any particular engine, the manufacturer's overhaul manual should be strictly adhered to.

The wide variations in temperature during the operation of the engine cause considerable expansion and contraction of its parts. Some of these distortions may cause permanent deformations which will make the reconditioning of the parts necessary after a certain number of hours of operation. Sludge chambers, corners, and joints in the engine structure will become filled with dirt, carbon, lead deposits, and sludge which will eventually make it necessary to disassemble the engine for cleaning.

Ring grooves and combustion chambers accumulate carbon deposits which should be removed. Such parts as piston rings and valves do not last indefinitely. Bearing surfaces may become scarred or scratched, and this damage, if allowed to become excessive, may lead to engine failure.

The length of time which an aircraft engine may operate before major reconditioning operations are necessary varies widely, depending upon the operating conditions. The usual period varies from approximately 350 to 800 hr. The Civil Aeronautics Administration defines two types of overhaul procedures: the top overhaul and the major overhaul.

Top Overhaul. The top overhaul consists of the removal, inspection, reconditioning, and replacement, when necessary, of the cylinders and piston assemblies. The propeller system, exhaust manifolds, valve-operating mechanism, ignition system, lubrication system, fuel system, cowling or cooling systems, heating systems, and the power-plant control systems are also included in the top overhaul. A top overhaul for any engine consists, in general, of the following procedure.

1. All cowling and inspection plates and panels should be removed

and inspected for defects. Cowling fastenings and chafing pads should be inspected. All necessary repairs and replacements should be made.

2. The propeller should be removed and inspected for its fit on the shaft, and marks, cracks, warping, distortions, or other defects on the blades. The propeller should always be removed when the cylinders are to be removed to prevent accidental turning of the crankshaft which might cause damage to engine parts.

3. The spark plugs should next be removed, cleaned, tested, and adjusted. The gaps should be set to the clearance recommended by the manufacturer.

4. The exhaust manifold and heating systems should be removed and inspected for leaks, condition of gaskets, and mounting attachments. Particular care should be taken to eliminate any leaks in the cabin heating system. Leaks in this system may cause carbon monoxide gas to enter the cabin and poison the occupants.

5. The intake manifold, or pipes, should be removed and inspected for dents, cracks, distortion, condition of gaskets, and proper mounting fixtures.

6. The cylinders and pistons should be removed in the order of rotation on a radial engine. The cylinder containing the master-rod piston should be removed last. If the master-rod cylinder is removed before the other cylinders, the entire assembly may turn, causing damage to the skirts of the other cylinders. As soon as a cylinder has been removed, its piston should be removed. It may be necessary to heat the piston-head with a blowtorch in order to remove the piston pins. Care should be taken that worn pins do not drop out when the cylinder is removed.

7. The oil sump should be removed and carefully inspected for metal chips, metallic particles, or any other foreign matter. *If metallic particles are found, the entire engine must be disassembled.*

8. The connecting rod and crankshaft assemblies should be given a careful visual inspection to determine whether or not there is abnormal wear. This inspection may be made through the cylinder base holes in the crankcase. Particular attention should be given to the proper safety-ing of all parts of the crankshaft and piston assembly.

9. The rocker arms and valve and spring assemblies should be removed from the cylinder. These parts should be carefully cleaned and inspected. All parts removed from the engine should be marked or arranged so that they may be easily identified and returned to their proper place in the engine.

Fig. 55. A cutaway end view of a 6-cylinder, inverted, in-line, aircraft engine. (Courtesy of Ranger Aircraft Engines)

10. The piston rings should be removed, and the pistons thoroughly cleaned of carbon inside and out. The lands between the grooves should be examined carefully for distortion, scarring, cracks, or wear. The old rings should not be replaced. New rings should, whenever possible, be installed.

11. The cylinder bores should be carefully inspected for scores, cracks, out-of-round, and proper taper. Cylinder wear may be measured by means of inside and outside micrometers or the proper dial gauges. The pistons should be measured to determine excessive wear. Go and no-go gauges should be used to check valve-guide clearance and wear. If the cylinders need grinding or honing, the manufacturer's recommendations should be carefully followed. These recommendations should govern the replacing of valve guides. If necessary, the proper oversize pistons should be used as replacements.

12. When the piston rings are installed in their respective grooves, care should be taken that they have the proper side clearance and end gap. The end gap may be measured in a ring gauge or by placing a ring inside the cylinder.

13. The piston pins and piston-pin bearings, as well as the entire piston, should be carefully inspected unless new pistons and pins are to be installed. Piston-pin retainers should usually be replaced but, if not replaced, the old retainers should be carefully examined to determine whether they are in good condition.

14. The piston pins should be thoroughly cleaned and measured for abnormal wear. Piston pins should be Magnafluxed to detect microscopic cracks.

15. Valve springs should be tested on a valve-spring tester and carefully examined for wear, rust, or cracks. Any visible defect should be reason for rejection of valve springs.

16. The rocker arms, bearings, rollers, and adjusting screw, as well as the rocker arm itself, should be carefully examined for signs of abnormal wear and defects. The ball end on the push rods should be carefully examined for excessive wear. Flat spots on the ball end or on the rocker arm roller should be cause for rejecting the part.

17. The valve seats and valve faces should be carefully refaced. Care should be taken not to leave any sharp or feather edges, as these tend to cause preignition. The seats should always be refaced after new guides have been installed.

18. The valve should be ground in by placing a grinding compound

between the valve face and the valve seat. In grinding or lapping valves, care should be taken not to rotate the valves continuously in one direction. The valve should be rotated back and forth. The grinding or lapping operation should be continued until there is a continuous matched surface completely around the valve face and the valve seat. All of the grinding compound should be carefully removed from all the parts with which it has come in contact. The thoroughly cleaned parts of the valve assembly should be reinstalled in the cylinder, and the valve ports filled with gasoline to check for any leakage between the valve face and its seat. If leakage occurs, rotate the valve slightly. If leakage continues, re-lap the valve.

Fig. 56. A general view of a vertical drive-shaft assembly on an inverted engine. (Courtesy Ranger Aircraft Engines)

19. Usually all gaskets which have been removed should be replaced with new gaskets during assembly. Use only the recommended compound on the gaskets being installed. Parts should be freely lubricated to assist in preventing wear when the engine is first started before the regular lubricating system becomes effective.

20. The master-rod piston assembly and cylinder should be the first to be assembled on radial engines. To install a cylinder, each piston should be in the top-dead-center position. The piston rings should be held in place with a piston-ring compressor while the piston is inserted in the cylinder. Check to see that all piston-pin retainers are in place. Care should be taken not to strike cylinder skirts on crankcase studs. Cylinder-base cap screws or nuts should be uniformly tightened, bringing all of the studs or nuts to the proper tension gradually. The nuts should be tightened with a proper torque wrench. After all nuts have been properly safetied or locked, the intake manifold or pipes should be installed.

21. The crankshaft should be turned until the piston in the No. 1 cylinder is slightly past the point where the intake valve closes. This will

place the piston in the top-dead-center position. The push rods and rocker arms should be thoroughly cleaned and installed, after which the intake and exhaust rocker arms are adjusted to the recommended cold clearance. This procedure should be followed for each cylinder in the firing order of the engine.

22. Before installing the propeller, check the seat on the hub if the engine is equipped with a tapered shaft. If there is any rocking or misfit, the hub may be lapped to a satisfactory fit by using a valve grinding compound. After the propeller has been installed, the retaining nut should be tightened and safetied. The tension, condition, and safetying of propeller hub bolts on wooden propellers should be carefully checked. On controllable or automatic-pitch propellers, the propeller control system should be carefully inspected and safetied. All miscellaneous items which have been removed should be replaced, properly secured, and safetied. It is always good practice to check the ignition timing to make sure that it is within the specified allowable limits.

Engine Major Overhaul. Major overhaul, as defined by the Civil Aeronautics Administration, consists of the complete reconditioning of the power plant and its various accessories. Some manufacturers of engines do not recommend the formal top overhaul. One or more cylinders may be removed, however, and reconditioned as described in the section on top overhaul.

It is not possible to give the detailed procedure of major overhaul for each make of engine. The following is more or less general in its scope, but these procedures may be applied to all aircraft engines.

General Overhaul Instructions. The major overhaul of an engine consists of the following operations.

1. The disassembly of the engine and all of its subassemblies into the individual parts
2. The thorough cleaning of all parts
3. The complete inspection of all parts
4. The repair, reconditioning, and replacement of worn or damaged parts
5. The complete reassembly of the engine.

The engine run-in and other tests after major overhaul may be included in the major overhaul operation. It is not possible to give detailed instructions for the major overhaul of all engines in any one book. The manufacturer's overhaul manual should be used when making a major overhaul on any engine.

It is important in order to perform a major overhaul on any engine that a satisfactory overhaul shop be available. The shop must be large enough to prevent crowding and must be free of dust, dirt, and excessive moisture. The shop should be well lighted and ventilated. The special tools required for any particular engine being overhauled should be available. The shop should be equipped with all the necessary general shop equipment.

Special inspection equipment, such as a Magnaflux machine, various types of micrometers, telescope gauges, and other necessary gauges and measuring tools, should be available.

A complete record system should be used, and the proper forms should be available to record the conditon of the engine upon its arrival at the shop. Forms should be used to record the condition of all parts of the engine as they progress through the overhaul shop. All inspection reports should be made on the proper form. Inspectors should

Fig. 57. An upper vertical drive assembly showing the relationship of the splined end of the shaft, bevel gear, and lower front idler gear of an inverted engine. Note the oil passage. (Courtesy Ranger Aircraft Engines)

check each operation as it is performed to prevent overlooking any important step in the overhaul of the engine.

After thoroughly cleaning the outside of the engine, the engine should be dismantled and disassembled. During the disassembling of the engine, all parts should be checked closely for signs of scoring, burning, excessive wear, and any other visible defect. This inspection should be made before the parts have been washed, as washing before careful examination may destroy some of the evidence which indicates the condition of the part.

The shop should be equipped with a proper lifting sling for the engine upon which the work is to be done. An engine assembly stand should be used. This stand should be of the rotating type which will allow the engine to be turned over without removing it from the stand.

A standard type of work bench and engine parts rack should be part of the equipment. The parts rack should be equipped with small racks, boxes, and containers in which to keep all parts of the engine separated and in their proper order. Unmarked major parts should be clearly numbered or tagged. The numbers or tags should indicate the exact location of the part on the engine and its position in relation to the other parts of the engine.

When removing nuts safetied with lock nuts (palnuts), always loosen and remove the lock nut before loosening the plain nut. Do not attempt to loosen or tighten the two nuts at the same time. All accessory pad openings, or any other opening, should be covered to prevent dirt or other foreign material from getting into the engine.

Antifriction Bearings. Antifriction bearings are made to very small tolerances, and special care should be used to prevent any foreign matter from entering the bearing. All such bearings should be wrapped in oiled paper as soon as they are removed from the engine. They should remain wrapped in this paper until they are to be cleaned. After these bearings have been cleaned, they should be oiled with a light engine oil. The bearings should be carefully inspected by a qualified inspector and not replaced unless the inspector recommends re-use. Antifriction bearings should never be inspected by the Magnaflux method. The iron powder in the liquid which enters the bearing can never be completely cleaned out. All antifriction bearings should be demagnetized at each overhaul period to prevent any small steel particles in the oil stream from being attracted by the bearing. Antifriction bearings cannot usually be reworked or repaired. A slight galling or roughness on the outside of the race may, however, sometimes be cleaned up by light stoning and polishing. After the bearing has been cleaned, inspected, and oiled, it should be wrapped in oiled paper until it is to be reinstalled in the engine. Usually this is done before demagnetizing.

Gears and Shafts. Most gears and shafts are subject to high stresses. All gears should be inspected with the greatest care. Parts of this kind should be Magnafluxed. Gear teeth and splines on shafts should be carefully examined for signs of roughness, pitting, or excessive wear. Small nicks or rough spots may be polished out by light stoning, if enough material is not removed to weaken the part. Usually, when one gear is discarded, the mating gear should also be discarded. Scratches on other steel parts, such as crankshafts, valves, and cylinder barrels, if not too serious, may be worked out by stoning. Two things must be considered when stoning a part.

1. Will the removal of the material affect the strength of the part?

2. Will the removal of the material change the shape of the part enough to cause increased wear on other parts?

Care should be taken in repairing parts to avoid making any sharp corners or abrupt changes in direction. Nicks and slight Magnaflux indications on shaft journals may be stoned out even to a point where a small flat spot is left on the journal. Such a flat spot should not fall at the edge of the bearing or it may cause a loss in oil pressure.

Castings. Nicks and scratches on machined surfaces of aluminum and magnesium alloy parts may be blended into adjacent areas by the use of a stone or a fine file. It is important always to smooth off any sharp corners. If the nick or crack extends across a parting surface, the surface should not be filed in such a way that a groove extends across the entire width because this would cause an oil leak.

When indications of corrosion are present on aluminum or magnesium alloy parts, the affected areas must be cleaned of all traces of corrosion. The rotary file is recommended to remove corrosion. Whenever a bare magnesium surface is exposed as a result of cutting, stoning, lapping, or filing, it must be protected at once by brushing with a sodium dichromate solution. (The proper method of preparing various solutions is given in Chapter IX, "Cleaning and Refinishing.")

Rough threads on a steel shaft may be smoothed with a file which is shaped to fit the groove of the thread. When the threads have been filed, the nut which fits on these threads should be lapped into place on the shaft.

Studs. Broken studs, dowels, or taps may be removed with a standard screw extractor. An Etchograph may be used. This machine burns out the broken end by means of an electric arc. When a defective stud has been removed from a casting, the original hole will be oversize. A standard stud should not be inserted in this hole. Studs are available in oversizes. These oversize studs are distinguished by having ends of different shapes. For example, a stud which is 0.003 in. oversize has a pointed end. A stud which is 0.006 in. oversize has a countersunk end, and a 0.009 in. stud has a hole drilled $\frac{3}{32}$ in. deep in the end. The original hole should be tapped to the correct size, and the new stud driven to the same depth as the old stud.

Reassembly. The most important consideration in reassembling an engine is to prevent any foreign object or dirt from entering the engine before or during assembly. All openings, such as cylinder holes in the crankcase, intake and exhaust parts and accessory drive openings, should be kept covered. Oilcloth bags may be used to place over cylinder holes in the crankcase. Cloth or paper should not be used to cover these openings because small pieces may tear off and fall into the engine. Wood-fiber board or stiff cardboard may be used.

Every part of the engine must be thoroughly cleaned. All moving parts must be coated with oil before being installed. After a part has been oiled and is ready for installation, it should not be touched again by cloth because of the possibility of threads clinging to the oily surface and thus entering the engine. All parts requiring safetying must be carefully safetied as soon as installed.

When nuts are placed on steel shafts, they should be screwed on by hand very slowly and, if the slightest binding is felt, they should be backed off at once. If a nut is forced over a bad thread, the

Fig. 58. The lower vertical drive-shaft assembly showing the bevel gear in mesh with the front camshaft gear. Note the oil passage. (Courtesy Ranger Aircraft Engines)

nut may seize and may cause the rejection of the part. If the thread appears rough, the nut should be lapped on the shaft with fine lapping compound. A light lapping compound furnished by the manufacturer or talcum powder and oil will usually correct slight roughness on the threads.

The proper thread lubricant should be used on all pipe plugs. The proper wrench should be used to install all plugs.

All gaskets, hoses, and leather parts should be replaced with new parts when reassembling the engine. Some engines use a silk thread as an oil seal between parting surfaces. The manufacturer's recommendation should always be followed in assembling the engine. Highly polished parts should never be touched by the hand or fingers after being cleaned and oiled. Salt from the perspiration or other foreign matter may start corrosion on highly polished surfaces. The assembly of an engine is one of the most critical duties of the aircraft-engine mechanic.

VIII DISASSEMBLY

Disassembly of the Engine into Its Major Subassemblies. For the complete disassembly of an engine, the mechanic should be furnished with the special tools recommended by the manufacturer for the particular engine upon which he is working. It is important that the proper tools should be used. Every manufacturer furnishes a list of special tools for his engine.

Before beginning to disassemble an engine, the outside of the engine should be cleaned of all traces of dirt and grease. This may be done by

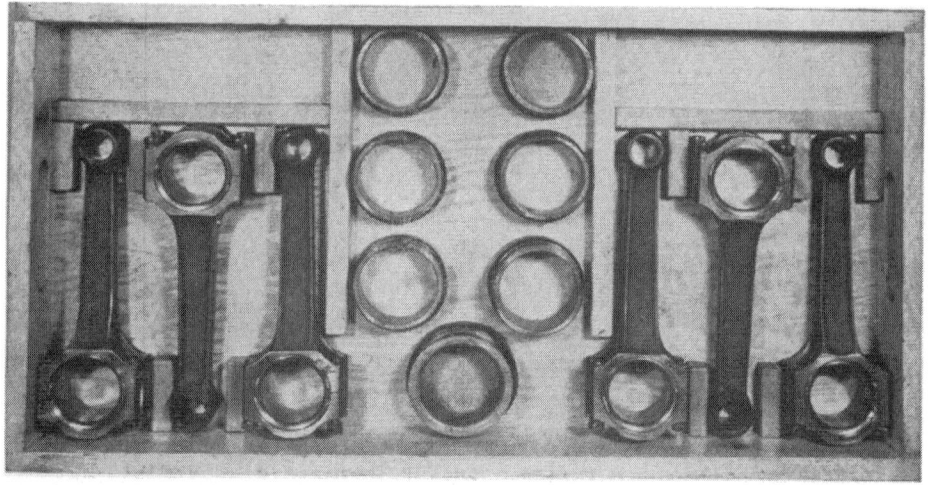

Fig. 59. Engine parts should be kept separate and in order. (Courtesy Ranger Aircraft Engines)

spraying the engine with unleaded gasoline, kerosene, or white furnace oil. It is important to have available suitable boxes, racks, containers, and benches so that the various parts or subassemblies may be placed in order as removed.

As each part or subassembly is removed from the engine, its general condition should be observed. The various movable parts, such as gears and shafts, should be tested for free or restricted movement. After the visual inspection, the subassemblies and parts should be thoroughly cleaned as they are removed from the engine. Before cleaning, each part or subassembly should be inspected carefully to note any unusual condition such as sludge or foreign matter deposits, particularly the pres-

Fig. 60. The engine is mounted on a stand before disassembly. (Courtesy **Ranger** Aircraft Engines)

ence of any metallic particles. If abnormal sludge deposits or metallic particles are found, samples of these deposits should be retained for careful examination.

Before beginning the disassembly of the engine, it should be mounted in a suitable rack, preferably one which allows the engine to be turned in various positions. After the engine is mounted in the rack, it should be locked in the desired position which, in the case of radial engines, is with the crankshaft pointing downward toward the floor. All cylinder baffles and baffle rings should be removed. The ignition wires should be disconnected from the spark plugs. The spark plugs should be removed and replaced with a set of vented dummy plugs.

The propeller shaft should always be checked with a runout dial indicator. The runout of the propeller shaft should be made at both the front and rear of the propeller cone seat with a dial indicator while the shaft is being turned with the special tool furnished for this purpose. If the runout exceeds the minimum allowed, a note should be made to check the shaft for possible bending when it is removed from the engine.

The thrust nut should next be loosened with the proper tool while the shaft is held stationary. If it is necessary to use a hammer on the

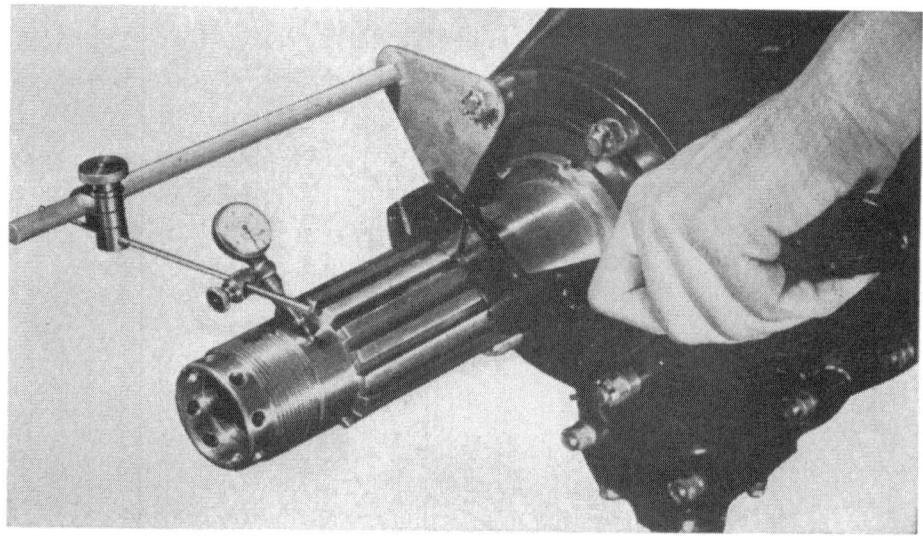

Fig. 61. Checking the crankshaft run-out at the propeller end. (Courtesy Ranger Aircraft Engines)

wrench to loosen the thrust nut, a lead hammer, or mallet, should be used.

The distributor blocks should then be removed from the magneto. Care should be taken that the molded block is not injured by its hitting the sides of the engine stand when removing the ignition manifold assembly.

The magneto, starter, fuel pump, generator, and accessory drive unit should then be removed.

Remove all nuts and clips securing the ignition manifold assembly, and remove the entire ignition manifold assembly.

Remove all oil drain hose lines connecting with the engine sump and allow the oil to drain. Loosen the hose clamps and clips on all external oil lines and remove all oil lines.

Remove the oil pump and oil-sump tube assembly.

The entire priming system should be loosened by removing the fittings in the intake ports of the manifold. Loosen the priming lines

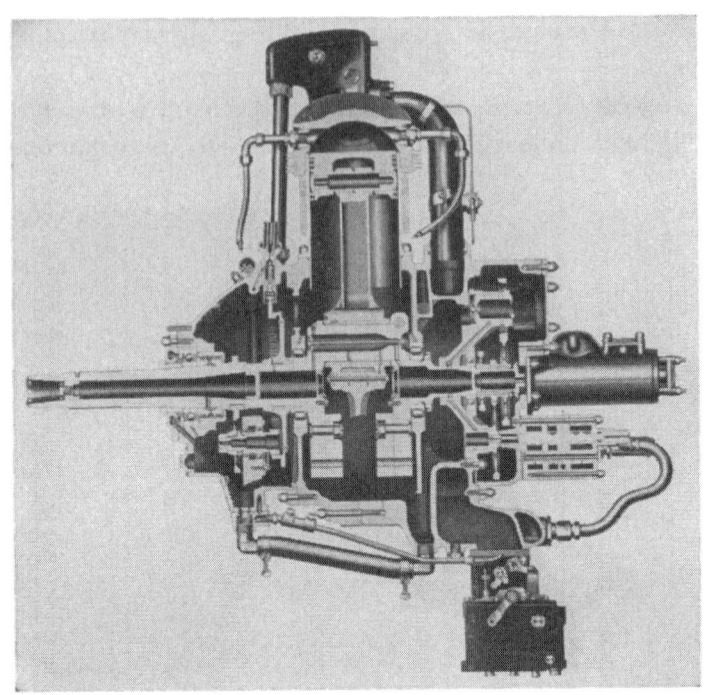

Fig. 62. A cutaway view of a 5-cylinder radial aircraft engine. (Courtesy Jacobs Aircraft Engine Company)

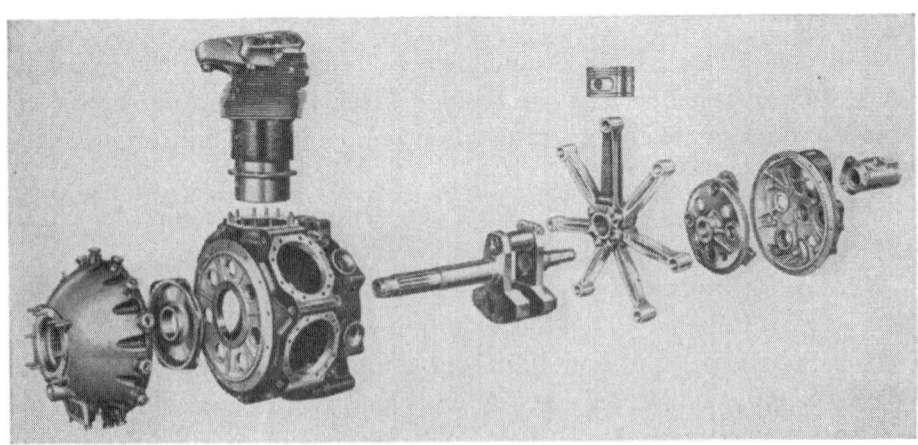

Fig. 63. An exploded view of a 5-cylinder radial aircraft engine. (Courtesy Jacobs Aircraft Engine Company)

from the priming distributor. Remove any clips which hold the priming lines in place. Remove all priming lines and tag each line to show the cylinder from which it was removed.

Remove attaching nuts and washers from the carburetor flange studs and remove the carburetor.

Remove the oil sump, fittings, oil-sump plug, and oil-sump strainers. Remove the oil-pump assembly.

Remove all nuts and washers which hold the rear crankcase and accessory gear covers in place. Remove the rear crankcase or accessory gear covers and remove the accessory gears using the special tools required. After the rear crankcase cover and accessory gears have been removed, radial engines should be turned so that the propeller shaft is upward and the rear of the engine is toward the floor.

All intake pipes should be removed, taking care that the intake pipes are not allowed to turn and jam the packing nut. Remove the intake

Fig. 64. Removing the intake pipe and manifold assembly. (Courtesy Ranger Aircraft Engines)

pipes by removing the intake port flange—attaching cap screws or nuts. Slide the flange over the intake pipes. After the intake pipes are removed, remove all packing.

The push-rod cover tubes, push rods, rocker-arm cover, and rocker arm should then be removed. Loosen the packing nut at the cylinder end first and push the nut over the cover tube. When using tools with an alligator type of jaw, knurled nuts should be protected by an aluminum or soft brass strip. Before removing the push rods, the crankshaft should be rotated until the rods are in their outer position. On most engines, the push rods are not all of the same length and should be marked or kept separated so that they may be put back into their

Fig. 65. Loosening the push-rod cover-tube packing nut. (Courtesy Jacobs Aircraft Engine Company)

Fig. 66. Removing the intermediate bearing plate. (Courtesy Jacobs Aircraft Engine Company)

proper place after each overhaul. When the push-rod cover tubes are fitted with a rubber washer, it will be necessary to break the rubber washer loose from the push-rod cover. The push-rod cover tubes can usually be removed by pushing each tube into the rocker box until the tube end clears the tappet guide. The tube may then be tipped away from the tappet guide and pulled out of the rocker box.

The cylinder, piston, and piston pins should be removed, leaving the No. 1 cylinder or master-rod cylinder until last. To remove any cylinder, rotate the engine until the piston in that cylinder is at top dead center. On engines equipped with automatic valve-gear lubrication, loosen the

Fig: 67. Removing a cylinder. The piston should be at top dead center. (Courtesy Ranger Aircraft Engines)

hose clamps on the hose connecting rocker boxes between each cylinder and remove the hose. Work around the engine, removing one cylinder after the other and being sure that the master rod is blocked in such a manner that the crankshaft cannot be accidentally turned. When removing the cylinder, care should be taken that worn piston pins do not drop out which would allow the piston and pin to fall to the floor with possible damage. After removal, the cylinder should be set down on clean wood or other soft surface to prevent damage to the skirt end of the barrel. In case the piston pins are difficult to remove, the piston-head may be heated by means of a blowtorch. Then the piston pin can usually be easily removed.

Fig. 68. Removing a piston pin using a plastic mallet and piston-pin drift. (Courtesy Ranger Aircraft Engines)

Fig. 69. Removing the front crankcase. (Courtesy Ranger Aircraft Engines)

The thrust nut, thrust bearing plate, and front section are then re-moved. The thrust nut is first removed, and then all of the attaching nuts and stud spacers holding the thrust plate to the front case. The thrust plate is then removed. If necessary, it may be tapped along the edge with a rubber mallet to loosen it.

The crankcase oil slinger and thrust-bearing-plate spacers should now be removed.

Remove all nuts and washers holding the front case to the crankcase. Before removing the front case, it is well to rotate the crankshaft three or four revolutions to make sure that all the tappets are moved to their outer positions. This is necessary in order that the tappet rollers may not catch on the cam lobes and damage the tappet rollers, pin, cams, or tappet guides. The front case is usually removed by means of a special puller.

Remove the timing gear spacer cam assembly and cam bearings. Re-

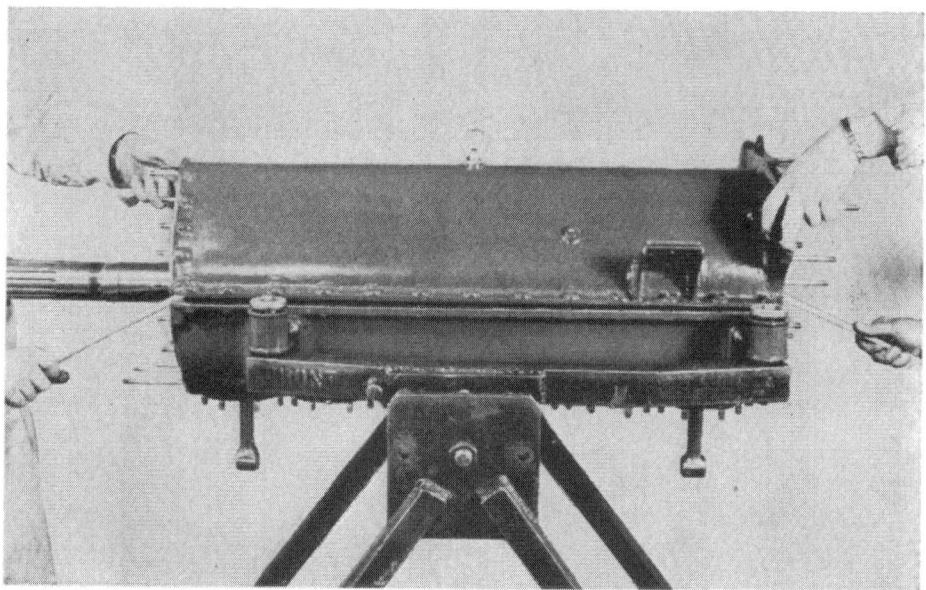

Fig. 70. Separating the halves of the crankcase. (Courtesy Ranger Aircraft Engines)

move the front half of the main crankcase, taking care not to damage the case.

The outer race or rollers of the front main bearing will usually re-main in the front half of the crankcase. They should be carefully re-moved to avoid dropping and should be kept separated to prevent mix-ing them with rollers from the other main bearings.

115

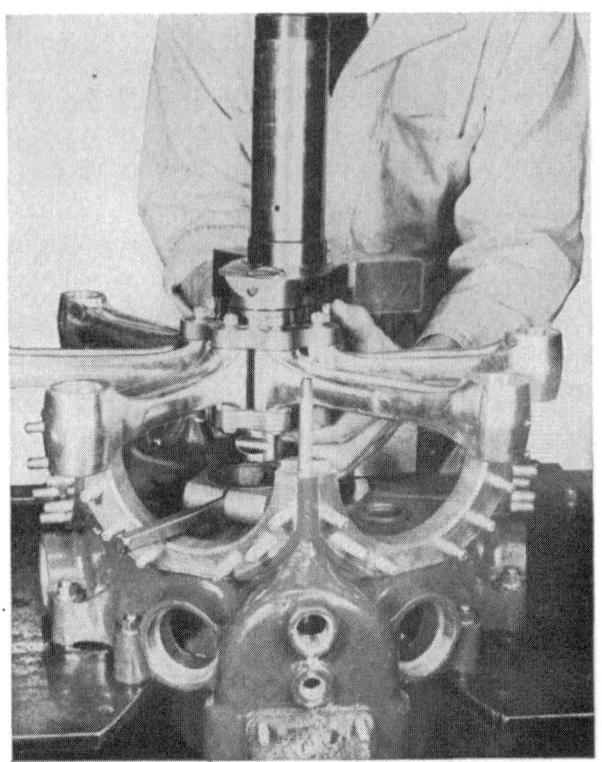

Fig. 71. Removing the front crankshaft and master and link-rod assembly. (Courtesy Jacobs Aircraft Engine Company)

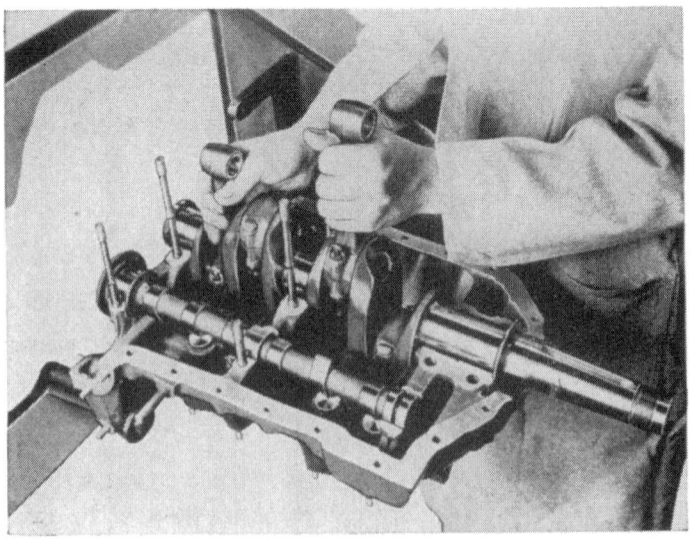

Fig. 72. Installing or removing a crankshaft assembly. (Courtesy Continental Aircraft Engines)

DISASSEMBLY

While the procedure described above pertains mainly to a radial engine, the same general procedure and precautions should be used with any other type of engine. With an in-line engine, the cylinders are usually lifted off vertically and may be removed in any order.

The crankshaft should be locked to prevent it from turning. The front crankshaft and master and link rods are next removed. If the crankshaft is in two parts, the crankshaft clamp bolt should be removed with the aid of a fiber drift. A special wedge is usually used to spread the opening slightly in the rear half of the crankshaft. Care should be taken not to damage the crankpin with the wedge. Carefully lift off the master-rod assembly together with the front half of the crankshaft. If the crankshaft is a single piece, the whole assembly is removed. Before removing the master-rod assembly from the crankshaft of the 2-piece type, carefully inspect the end of the crankpin for any burrs or rough spots. Any roughness on the end of the crankpin will scar the master-rod bearing when it is removed. Any burrs or rough spots should be carefully removed before attempting to remove the master-rod bearing. The master-rod and link-rod assembly may then be removed from the crankpin. The engine is then usually rotated to the horizontal position before removing the rear crankshaft.

The rear half of the crankshaft is removed together with the rear main bearing. Care should be taken to avoid dropping these bearings.

Fig. 73. Removing a rear-cover assembly. (Courtesy Kinner Motors, Incorporated)

117

A special puller may be necessary to remove the inner races. Care should be taken not to mix the rear roller bearings with the front roller bearings. Recently manufactured main roller bearings are serially numbered on the inner and outer races to avoid mixing, but the rollers

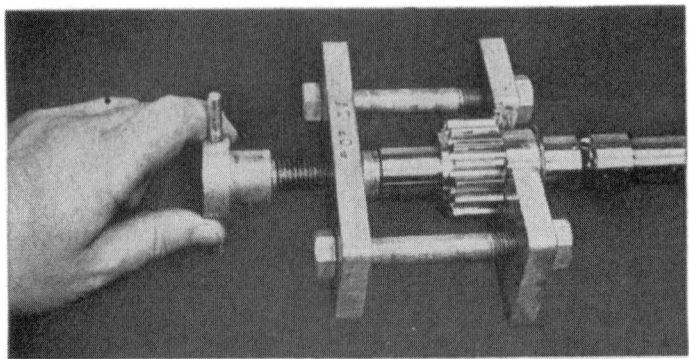

Fig. 74. Removing an accessory drive-shaft gear. (Courtesy Ranger Aircraft Engines)

Fig. 75. Removing a magneto drive gear. (Courtesy Ranger Aircraft Engines)

should be kept separate, anyway. After the front and rear halves of a crankshaft are removed, it is advisable to place some suitable protection over the crankpin. A thread protector should be screwed on the end of the front half in order to protect the threads.

DISASSEMBLY

Disassembly of Major Subassemblies. At each engine overhaul, the ignition manifold assembly should be completely disassembled and all ignition cable replaced. The ignition manifold should be disassembled by first removing the ignition cable markers and distributor blocks.

Fig. 76. Removing a piston-pin circlip. (Courtesy Ranger Aircraft Engines)

Fig. 77. Removing piston rings. (Courtesy Jacobs Aircraft Engine Company)

The distributor and conduit assemblies, coil conduit assembly, spark-plug terminals on the end of the spark-plug cables, spark-plug elbows, and spark-plug conduit are all removed. The wires can then be pulled from the manifold.

The oil pump should be completely disassembled. The bolts holding the various sections of the oil pump together should be removed, and

Fig. 78. The piston rings are removed by means of a piston-ring expander. (Courtesy Ranger Aircraft Engines)

the oil-pressure-relief valves should be disassembled. Most oil pumps contain both a pressure unit and a scavenger unit. The oil-check-valve spring and ball should be removed from the pressure section of the pump.

On engines having automatic valve-gear lubrication, there will be an extra section called the rocker scavenger section. This section will also have to be disassembled.

The rear case assembly and rear intermediate bearing-plate assembly should be completely disassembled.

Next the cylinder and pistons should be disassembled. To disassemble a cylinder, it should be placed on a suitable block for support. The rocker-arm shaft nut should be removed. The rocker-arm shaft should be tapped out, and the rocker arm removed.

Before attempting to compress the springs with the valve-spring compressor, the upper valve-spring washer should be tapped to loosen the washer from the locks and prevent straining the compressor. After the springs are compressed, remove the locks. After the compressor is removed, lift the cylinder from the block holding the valves so that they do not fall out. The valves are then removed and marked or numbered so that they may be returned to their proper place. When removing the rings from the pistons, care should be taken not to scratch or raise any

Fig. 79. Compressing the valve spring by means of a valve-spring compressor. (Courtesy Ranger Aircraft Engines)

burrs on the lands between the grooves or on the sides of the pistons. Rings are usually replaced at the overhaul period. If the rings are not to be replaced, care should be taken to avoid twisting or expanding them more than is necessary to remove them from the groove. The rings should be numbered and kept in exact order and each replaced in the same groove from which it was removed.

The front case should be completely disassembled and it is usually turned over and allowed to rest on the cowling studs. The various parts should be removed with the tools provided for that particular operation.

The front and rear crankshafts should be disassembled by removing the crankpin bolt nut, crankpin bolt, and crankpin plug. To remove the rear crankshaft oil plug, it is necessary to drill the plug at the points where it is staked. As small a drill as possible should be used to remove the punch mark. Do not drill deeper than necessary. After drilling, the plug is removed with a special wrench. The staking method of locking the crankshaft plugs has been replaced in most engines by

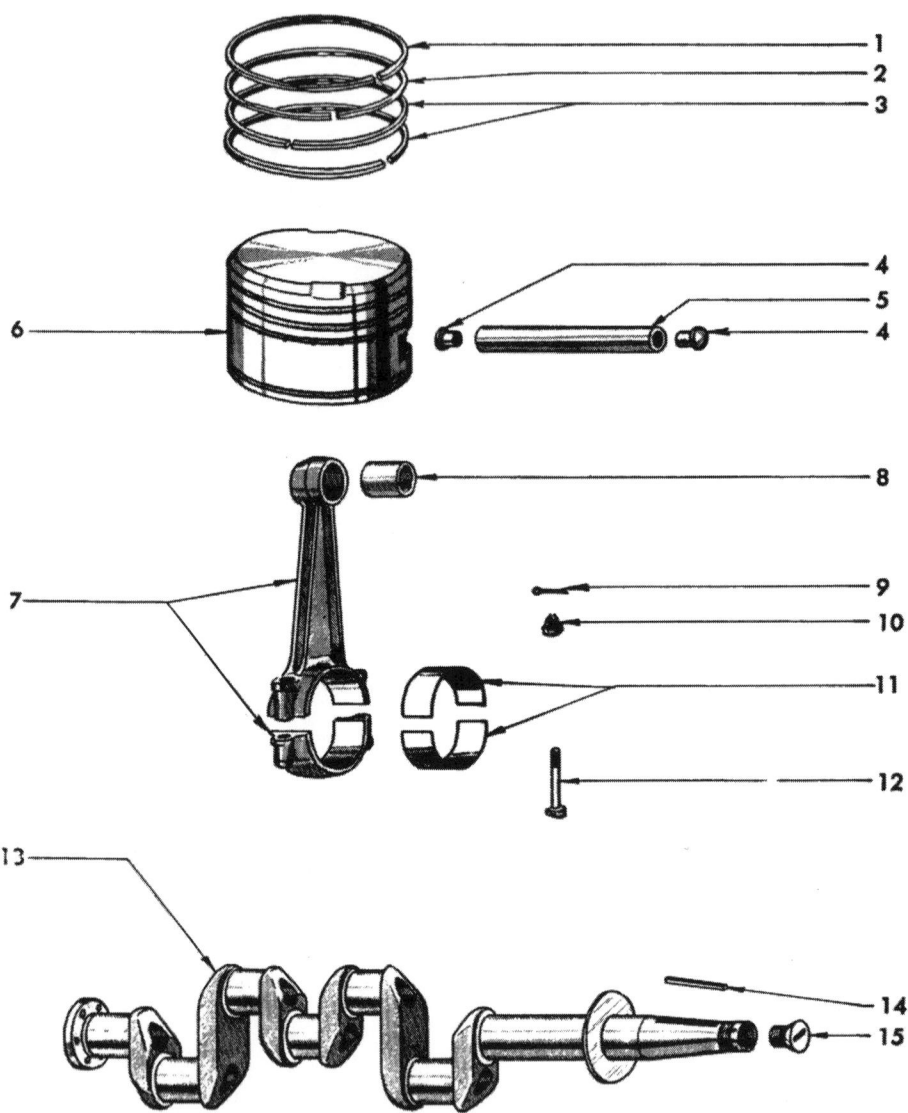

Fig. 80. The crankshaft, a connecting rod and piston for a light 4-cylinder aircraft engine: (1) Piston-compression ring; (2) plain compression ring; (3) oil-control piston ring; (4) piston plugs; (5) piston pin; (6) piston; (7) connecting-rod assembly; (8) piston-pin bushing; (9) cotter pin; (10) castle nut; (11) upper and lower connecting-rod bushing; (12) connecting-rod bolt; (13) crankshaft assembly; (14) propeller hub key; (15) front-end-crankshaft oil plug. (Courtesy Continental Aircraft Engines)

locking the plugs with a lock and lock screw which are safety wired. To remove plugs locked by this method, it is only necessary to remove the safety wire, lock, and lock screw. Plugs can then be removed with special wrenches.

To disassemble a master and link rod assembly, the cotter pins and

Fig. 81. Removing connecting-rod bolts. (Courtesy Ranger Aircraft Engines)

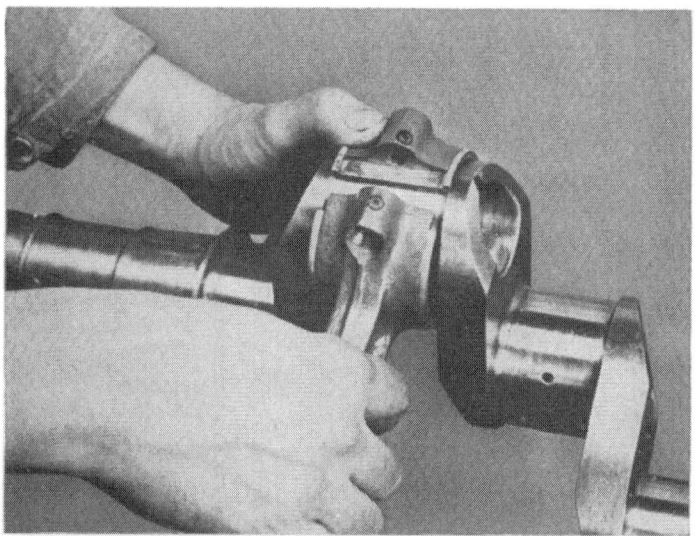

Fig. 82. Removing the connecting rod from a crankshaft. (Courtesy Ranger Aircraft Engines)

other safeting devices should first be removed. The knuckle pins usually have to be pressed out. Before pressing out the knuckle pins, the master rod should be placed on a special base. The small, drilled oil holes inside the master-rod bearing should be down and the bearing lock plate upward. The oil-groove end of the knuckle pin will then be

Fig. 83. Removing the crankshaft oil plugs from the main journals. (Courtesy Ranger Aircraft Engines)

Fig. 84. Removing the large crankshaft plugs using a puller. (Courtesy Ranger Aircraft Engines)

down. This end should be pressed out first. If the knuckle pin is pressed out from the wrong side, the groove may pick up metal and scar the hole in the master rod. The manufacturer's recommendations should be carefully followed when removing the knuckle pin to avoid damage to the master-rod flanges.

Fig. 85. Removing the hollow dowel pins in the crank cheek. (Courtesy Ranger Aircraft Engines)

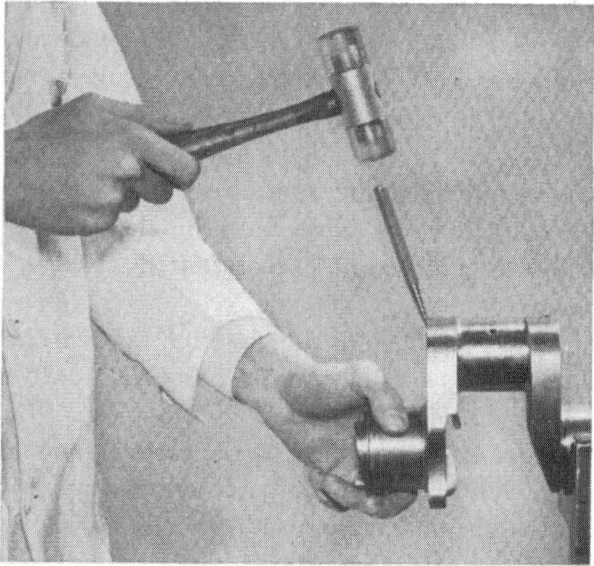

Fig. 86. Removing the pin locating the balancing plug in the No. 1 crank pin. (Courtesy Ranger Aircraft Engines)

125

IX CLEANING AND REFINISHING

Cleaning. All parts of the disassembled engine should be carefully cleaned. The usual practice is to degrease all parts of the engine as soon as the disassembly has been completed. The usual degreasing methods are quite effective in removing soft-sludge carbon deposits. Hard carbon deposits, such as are found on the piston assembly and in the combustion chamber, require a decarbonizing process. All parts made of magnesium should be separated from the steel and aluminum parts. Magnesium parts must be cleaned with a neutral, noncorrosive degreasing compound. Great care should be used to prevent fire when washing parts with unleaded gasoline spray, white furnace oil, or kerosene. Safety naphtha is recommended.

It is difficult to remove from engine parts all traces of water-mixed cleaning solutions which contain caustic compounds or soap. When these compounds are used, oil foaming may result immediately after starting the engine, or within a few hours. The alkaline compounds combined with the oil in the engine form soaps which produce oil foaming. If water-mixed cleaning solutions which contain soap or caustic compounds are used, the parts must be thoroughly scrubbed in boiling water and then rinsed in a separate bath of clear boiling water after using them.

It is a much better practice to use gasoline, kerosene, or some other hydrocarbon liquid for cleaning. Leaded gasoline should never be used for cleaning. Grease and soft carbon compounds may be removed by spraying with white furnace oil, kerosene, carbon tetrachloride, safety naphtha, or unleaded gasoline. The parts being cleaned may either be dipped into the solution, or sprayed or brushed with it. Cleaning should always be done in a semienclosed, well-ventilated booth or hood. As soon as the parts are clean, they should be sprayed with a light lubricating oil to prevent rust or corrosion.

Hard carbon deposits must be removed with a decarbonizing solution. Most decarbonizing solutions will also remove enamel from engine parts. Some of these preparations will attack aluminum and magnesium parts if they are allowed to stand too long in the solution, however,

Fig. 87. Cleaning an engine part by means of a solvent spray. (Courtesy Ranger Aircraft Engines)

and the mechanic should follow the manufacturer's directions carefully.

After the external surfaces of the parts are clean, special care should be taken to see that all internal passages, such as built-in oil passages, are thoroughly cleaned and blown out with compressed air. The openings of these passages should be covered with Scotch tape or closed by some other method to prevent dirt from entering before the parts are reassembled.

The pistons should be placed in a cleaning solution until the carbon has been softened enough to be wiped off or removed with a soft metal or wooden scraper. Particular care should be taken in cleaning the ring grooves. Every precaution should be used to prevent damage to the lands or the removal of any aluminum from the small radius where the lands join the bottoms of the ring grooves. It is often necessary to sandblast the inside surfaces of the pistons to remove the carbon. Sandblasting may be necessary to detect any cracks on the inside of the piston. However, sandblasting should not be done until after the piston-

Fig. 88. Cleaning inside of a crankcase with a solvent spray. (Courtesy Ranger Aircraft Engines)

pin holes have been plugged. The outside of the piston should be covered from the bottom of the skirt to the top of the head with rubber tubing or another protective material. Glazed surfaces on the piston skirts should not be removed. Oil-relief holes may be cleaned of carbon by reaming with an undersized drill. Final cleaning of the piston may require polishing with crocus cloth and kerosene, followed by a final spray washing. Hard carbon within the combustion chamber and enamel or paint from the outside of the cylinder may be removed by sandblasting.

When sandblasting cylinders, the same precautions as those given for pistons should be followed. All openings and polished surfaces must be protected. The inside of the cylinder barrel must be protected by a suitable cylinder-wall protector. Rubber plugs should be placed in the valve guides from the inside. Threads in the spark-plug bushing should be protected with rubber plugs or with a set of dummy, or discarded, spark plugs.

The valve seats may be sandblasted. This operation often cuts the carbon or glaze from the seat. This assists reconditioning, particularly with the exhaust-valve seat.

It is not always necessary to remove all the enamel from the outside of the cylinder. If the enamel is to be removed by sandblasting, all openings must be plugged. The flanges and studs should also be pro-

Fig. 89. Cleaning a cylinder with a solvent spray. (Courtesy Ranger Aircraft Engines)

Fig. 90. Machined surfaces must be protected before sandblasting or finishing. (Courtesy Ranger Aircraft Engines)

tected. After sandblasting, the sand should be removed by the use of compressed air and gasoline. Particular care should be taken in removing any sand which may lodge in the joint between the cylinder head and the cylinder barrel. Any hard carbon remaining on the valve heads after they are removed from the cleaning solution may be removed with a fine wire brush.

Extreme care must be used in cleaning the master rod and its bearings. Cleaning solutions have a tendency to remove lead particles. If lead is used in the surface of the bearing, the bearing should be cleaned by carefully wiping it with a clean cloth wet with unleaded gasoline or carbon tetrachloride. Polishing, burnishing, or subjecting the bearing to any other cleaning solution except unleaded gasoline or carbon tetrachloride should be avoided.

Roller and ball bearings should be cleaned with a neutral cleaning solution. Hard carbon may be removed from bearings by placing them in a carbon solvent which will not attack the bearing or the retainer. Most bearing trouble is caused by dirt getting into the bearing. The cleaning of a bearing is a particularly painstaking job, and should be performed by a mechanic who completely understands this work and who can carry it out with thoroughness and caution.

Antifriction bearings are subject to magnetization. These bearings may become magnetized while in transit, in storage, or in the engine. Small particles of magnetic material will stick to the balls and races

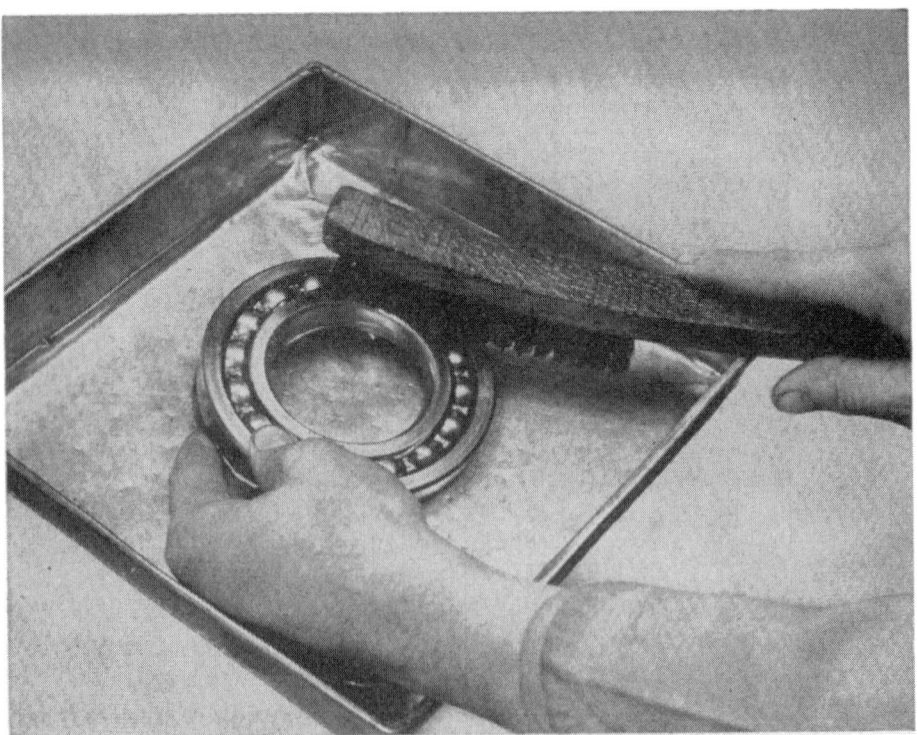

Fig. 91. Cleaning ball bearings with a solvent and soft brush. (Courtesy Ranger Aircraft Engines)

when they are magnetized. These particles will cause scoring of the parts during operation. It is recommended that all bearings be demagnetized by passing them through a demagnetizer. The bearings should then be thoroughly washed and oiled before installing them in the engine. New bearings should always be passed through a demagnetizer while still packed in the original container.

After being removed from a cleaning solution, the bearings should be washed and blown out with compressed air. It is important that the compressed air be free from all particles of dirt and from moisture. The bearings should not be allowed to spin while being blown out. The spinning of the bearing while dry will cause scoring of the races. After cleaning, the bearings should be oiled, wrapped in waxed paper, and placed in a box to prevent dirt from getting into the bearing. A clean bearing should never be allowed to remain exposed on a workbench.

After cleaning, unless they are to be inspected immediately, all steel parts must be covered with a coat of rust-preventive oil. This may be

done either by spraying or dipping. Under no conditions should a steel part be allowed to remain dry for more than 2 hr. Engine parts should not be allowed to remain in contact with each other in the engine-parts rack.

Refinishing. Most of the external parts of an aircraft engine must be refinished at major overhaul periods.

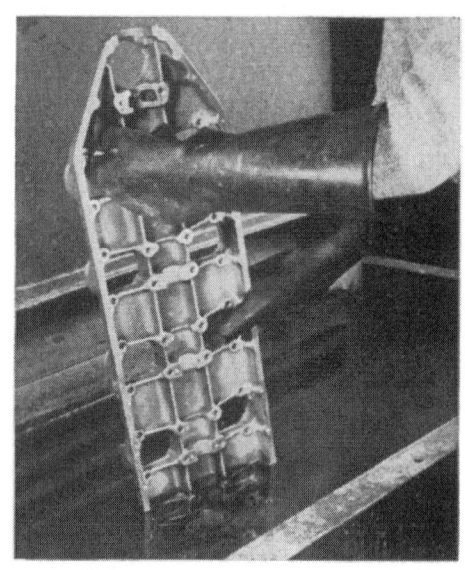

Finishing Steel Parts. If the part needs to be refinished by enameling, the following procedure may be followed.

1. Wash the part in a mixture of gasoline and carbon tetrachloride.

Fig. 92. Parts must be thoroughly cleaned before refinishing. (Courtesy Ranger Aircraft Engines)

2. Sandpaper the original enamel until a smooth surface is obtained. Sandpapering is particularly necessary on chipped or rough surfaces in order to properly smooth them.

3. Dust the part with dry compressed air.

4. Protect all machined surfaces with special covers or by the use of masking tape.

5. Spray on the enamel recommended by the manufacturer.

6. Most enamel should be baked on by placing in a flameless oven at a temperature of approximately 250° F. Parts should remain in the oven for approximately 2 hr.

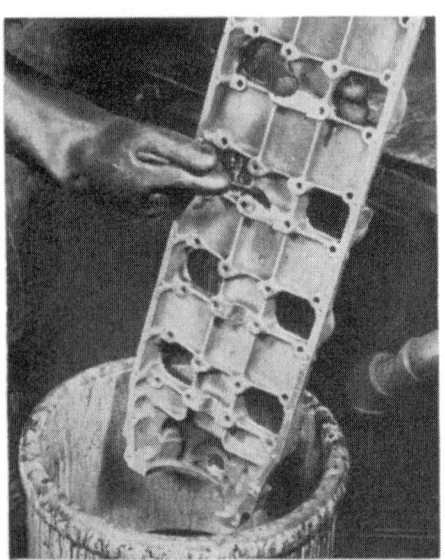

Fig. 93. Applying cold sodium dichromate solution to a magnesium alloy casting. This solution will affect cadmium plating. (Courtesy Ranger Aircraft Engines)

Finishing Magnesium Parts. If parts made from magnesium alloy need refinishing, the following procedure may be used.

1. Remove the old enamel by sandpapering or dipping in a paint-remover solution. Care must be taken not to leave parts of this alloy in the solution long enough to allow the solution to attack the metal.

2. Remove all traces of dirt by cleaning with a solvent spray.

3. Wash the part thoroughly in hot running water.

4. Apply cold sodium dichromate solution with a brush. Care should be taken to prevent this solution from touching the studs or other parts as it will affect cadmium plating.

5. Rinse in cold water and then dip in hot water.

6. Dry rapidly with compressed air.

7. Apply one coat of zinc chromate with a spray gun and air dry.

8. Spray on three coats of the recommended enamel, baking each coat in a flameless oven for approximately 1½ hr. at a temperature of 300° F.

Finishing Aluminum Parts. When aluminum parts require refinishing, the following procedure may be used.

1. Remove the original finish by sandpapering or dipping in a paint remover. Aluminum parts should not be left in paint remover long enough so that the remover will attack the metal.

2. Clean off all traces of dirt with a solvent spray.

3. Wash thoroughly in hot running water.

4. Dry with compressed air.

5. Spray on one coat of zinc chromate primer and air dry.

6. Spray two coats of the recommended enamel, baking on each coat for about 2 hr. in a flameless oven having a temperature of about 280° F.

Cadmium Plating. All exposed parts made of steel or brass that are not otherwise finished should be cadmium plated as a protection against corrosion. The plating is normally less than 0.001 in. in thickness. Cadmium is quite soft, and care must be used not to scratch through this thin plate during overhaul or servicing of the engine. To apply cadmium plate, the following procedure may be followed in the order given:

1. All traces of grease should be removed from the parts to be plated by placing them in a wire tray above a solution of trichloroethylene. The liquid is heated by steam pipes, and the vapor formed rises past the parts and removes the grease. The vapor should be condensed by cold water pipes running through the top of the tank above the parts being treated. The condensed liquid runs back into the tank and is saved.

2. Strip the part by dipping in a solution of ammonium nitrate or hydrochloric acid for about 30 min.

3. Soak the parts in a solution of sodium hydroxide (caustic soda) for about 30 min. (The formula for this mixture is given at the end of this chapter.)

4. Place all parts in a tank containing the cadmium-plating solution given at the end of this chapter. Large parts will have to be suspended separately on wires. Small parts should be placed in a tumbler to keep the parts in motion during the plating process. If the small parts are not kept in motion, the plating will not be evenly applied.

5. The plating is done by means of a direct current of electricity. The parts to be plated are connected to the negative terminal of the current source. Bars of pure cadmium are suspended in the solution connected to the positive terminal of the current source. The amount of current depends upon the size of the tank and the number and size of the parts to be plated.

6. After removal from the plating tank, the parts should be bright-dipped in water containing 1 per cent nitric acid. After bright-dipping wash thoroughly in clean cold water.

Sandblasting and Metalizing. Many manufacturers recommend that cylinders be remetalized at each overhaul. Cylinders or other parts to be metalized should be degreased and sandblasted. The parts to be metalized should be degreased with trichloroethylene vapors. As soon as the cylinders are completely cleaned and dry, a coat of rust-preventive oil should be applied to the inside of the cylinder barrels. Care should be taken to see that no oil comes into contact with the surface to be

Fig. 94. Before metalizing, cylinders must be degreased and then sandblasted. Complete degreasing is done with trichloroethylene vapor. (Courtesy Ranger Aircraft Engines)

metalized. When sandblasting, the parting surfaces and the inside of the barrel should be carefully protected with proper covers. The quality of the metalizing finish depends largely upon the care with which the sandblasting was done.

Air at a pressure of about 85 lb. per sq. in. should be used for at least 5 min. in cleaning each cylinder assembly. After the cylinders are sandblasted and cleaned, they should be lifted only by the masking equipment, if possible. Parts to be metalized should be handled only with clean cotton gloves. The metalizing will not cling to any surface which has been touched by the exposed skin. The metalizing should be done immediately after cleaning because moisture on the surface will prevent the molten metal's sticking.

Metalizing is a method of spraying metal with molten metal from a metal spray gun. The metal in the form of a wire is fed through the spray gun and melted in a flame of oxygen and acetylene. The melted metal is sprayed onto the surface to be metalized by air pressure. The air pressure should be approximately 60 to 65 lb. per sq. in. The oxygen

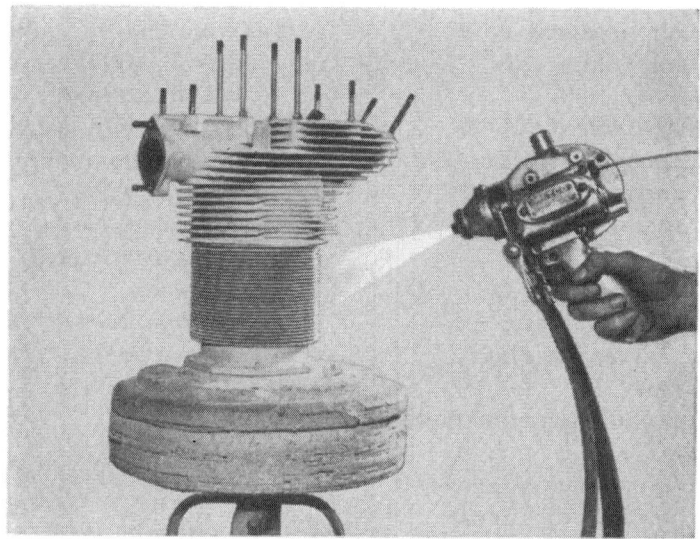

Fig. 95. Metalizing the cylinder surface by spraying with molten aluminum from a metal spray gun. (Courtesy Ranger Aircraft Engines)

should be furnished at 17 lb. per sq. in. and the acetylene at a pressure of 16 lb. per sq. in.

Metalizing is done on parts most effectively when the part is mounted on a turntable which rotates the part as the spraying is being done. The metalized coat applied to cylinders is of aluminum. Any surface which is at right angles to melted aluminum particles will have the required coat built up rapidly. The thickness of the metalized coat is from 0.005 in. to 0.006 in. If the metal spray strikes the surface at an angle, the coat will be correspondingly thinner. Corrosion on cylinders occurs most frequently between the fins of the cylinder barrel, and care should be taken to spray the fin at an angle so that the sides of the fins are covered with an even coating of the aluminum. The spray should strike the fins at an angle of about 30° from above the fin and at an angle of about 30° below. Spray one side of the fins completely before spraying the opposite side. Much of the success of a major overhaul depends upon the care with which the cleaning and refinishing has been done.

CHEMICAL FORMULAS USED IN ENGINE CLEANING
AND REFINISHING

Honing Coolant
  15 gal. Almag
  1 gal. International
    Compound #155

Grinding Coolant
  4 gal. International
    Compound #137
  75 gal. water

Valve-Facing Coolant
  ½ gal. Economy Lubricant E–1
  5 gal. water

Exhaust-Valve-Seat Grinding Coolant
  7 parts Mineral Seal Oil
  1 part Cut-Max. Baste #2

Magnaflux Inspection Fluid
  98% Varsol
  2% iron oxide (paste or powder)

Dichromate Solution
  7½ qt. nitric acid
  13 lb. sodium dichromate
  32 qt. water

Caustic Soda Solution
  85 lb. caustic soda
  18 lb. trisodium phosphate
  18 lb. sodium cyanide
  150 gal. water

Cadmium-Plating Solution
  118.5 lb. sodium cyanide
  28.1 lb. cadmium oxide
  12.2 lb. sodium hydroxide
  28.1 lb. ammonium sulphate
  150 gal. water

Muriatic Acid Mixture
  50% muriatic acid
  50% water

Ammonium Nitrate Mixture
  20% ammonium nitrate
  80% water

X OVERHAUL INSPECTION

Inspection. Inspection is the most important part of the mechanic's job and requires more experience and better judgment than any other single item. The inspector should thoroughly understand the tables of limits which are prepared by each manufacturer for his particular engine. The inspector must have the ability to decide what is wrong with any part of the engine and whether or not parts should be reused, repaired and reused, or discarded. A typical table of limits is given at the end of this chapter. As soon as the parts of the disassembled engine are degreased, they should be carefully inspected.

The quality of inspection depends directly upon the experience and judgment of the person performing this operation. Inexperienced mechanics or other personnel should never be allowed to make an inspection without the supervision or instruction of a thoroughly experienced inspector.

The inspection should include not only visual inspection but tests and measurements of all working fits and clearances. These fits and clearances may be found by referring to clearance charts and tables and the table of limits for each particular engine. The clearance chart and table of limits give the minimum and maximum limits of the desired fit at each location, and the replacement limit. The tables also give the limits for spring loads. The inspector should make a record of wear and clearances on major engine parts. The inspection report should include notes pertaining to any unusual condition found in the engine.

Parts requiring replacement or repair should be plainly tagged by the inspector. Any work to be performed on a part should be noted on this tag.

All cast parts should be carefully examined with a magnifying glass before the paint is removed. A crack often shows up plainly through the paint, as the paint itself will be cracked. All parts of the engine should be neatly arranged on an inspection table which is large enough to hold one complete engine. Visual inspection should, of course, always be given parts during disassembly. This inspection is made before the parts are degreased. Many defects are plainly visible at this time.

Fig. 96. Using a micrometer to check the diameter of the main bearing journals and crankpins. (Courtesy Ranger Aircraft Engines)

Fig. 97. Many engine parts must be examined with a magnifying glass to detect defects. (Courtesy Ranger Aircraft Engines)

All steel parts, except such parts as antifriction bearings, nuts, washers, cotter pins, and low-stress parts, should be Magnafluxed. Magnafluxing is a process during which the part being inspected is clamped in the Magnaflux machine and a heavy current of low-voltage electricity is passed through the part. While under the influence of the electric current, the part is treated by being dipped into, or having poured over it a solution of Magnaflux. This Magnaflux solution contains a finely divided magnetic material. Any crack, blemish, or inclusion which causes a break in the surface of the part becomes a small magnet. The magnetic material will cling to the opposite poles of these small magnets, and microscopic failures may be readily discovered.

Magnafluxing should be done by thoroughly experienced personnel and the instructions of the manufacturer of the Magnaflux equipment should be carefully followed.

All bearings should be carefully inspected, and bearings that cannot be disassembled should be inspected visually for pitting, galling, flaking, and excessive wear of balls or races. Bearings should be checked for roughness by holding the outer race and rotating the inner race while exerting side pressure. This pressure should be exerted in both directions to check both sides of the bearing. Radial pressure should be ex-

Fig. 98. Checking the run-out of a bearing race. (Courtesy Ranger Aircraft Engines)

erted on the inner race to check for any roughness. Slight roughness or wear may or may not be a cause for rejection. If close visual inspection reveals pits or any other defect on the races or balls, the bearing should be rejected.

Bearings which can be disassembled should have the roller and races carefully checked for pitting, galling, flaking, and excessive wear. The bearings should be inspected for the condition of the inner diameter of the inner race and the outer diameter of the outer race of the front and rear surfaces. If the surfaces are found to be slightly galled or scored, the bearing should be tagged for repair by stoning.

All studs should be checked and inspected for straightness, tightness, stretching, and damage to the threads. Any studs which show signs of stretching should be tagged for replacement. Studs showing other damage should be replaced unless it is evident that they can be easily repaired.

All liners or bushings should be checked for looseness. The condition of their bores should be carefully inspected for pitting, scoring, galling,

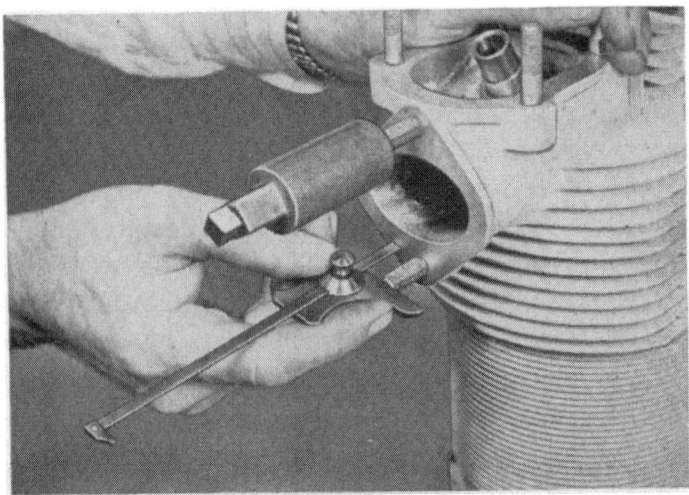

Fig. 99. Measuring the depth to which a stud has been driven. (Courtesy Ranger Aircraft Engines)

Fig. 100. Determining the alignment of the intake port by the use of a feeler gauge. (Courtesy Ranger Aircraft Engines)

or any other unusual wear. Bushings and liners that are loose or show excessive wear should be replaced.

All cast parts of the engine, such as crankcase sections, should be examined for nicks, cracks, and other signs of damage or indications of failure. All capped holes should be examined for thread damage. The condition of all flange faces and mating surfaces should be checked for indications of corrosion or any indication of poor fit.

Magnesium parts which show signs of corrosion but are in a repairable condition should be tagged for an application of the dichromate pickle treatment described on page 132.

All passages, particularly drilled oil passages, should be checked to make sure that they are clean and open. As the extent of a fine crack on a casting cannot be easily traced, the surface of the casting should be thoroughly cleaned. An agitated solution of azure-blue powdered chalk in acetone should be prepared and sprayed on the casting. The traces of oil which remain in the crack will show as a heavy dark line after thorough drying. Azure-blue powdered chalk may be secured from the manufacturer.

Gears and shaftings should be carefully checked for cracks, galling, wear, and mutilation. Gear teeth may be slightly uneven, but need not

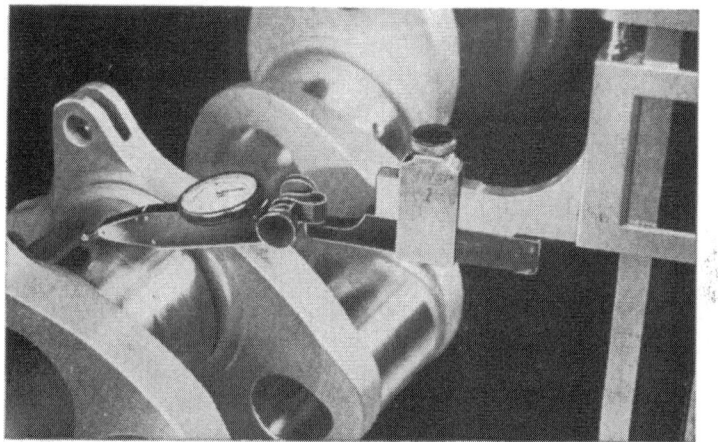

Fig. 101. Checking the run-out of the crankshaft bearing. (Courtesy Ranger Aircraft Engines)

be rejected for this reason. Slightly worn or pitted gears may usually be reinstalled providing the backlash does not exceed that listed in the table of limits. Shafts should be checked for straightness. Straightness may be checked by mounting the ends of the shaft on V-blocks or in a lathe. No attempt to straighten shafts should ever be made. Bent shafts must always be replaced.

The crankshaft master-rod assembly, or connecting-rod assembly, should be examined to make sure that all parts are thoroughly clean. This applies particularly to oil passages and sludge chambers. Hollow crankshafts, in particular, require careful examination. Splines and propeller mounting seats should be carefully examined for nicks, scoring, wear, or cracks.

Fig. 102. Checking the run-out of a main journal with respect to other main journals. (Courtesy Ranger Aircraft Engines)

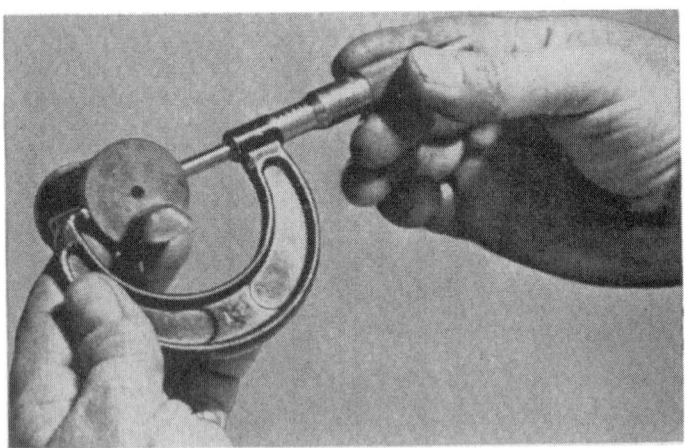

Fig. 103. The crankshaft plugs should have their diameters checked by the use of a micrometer. (Courtesy Ranger Aircraft Engines)

Keyways and keys should be checked for burrs and scores. All threads should be inspected for condition. Check all internal crankshaft plugs for proper position. Thrust-bearing seats and rear cone seats should be examined for galling. The crankpins should be carefully examined for scoring, burrs, wear, or unevenness. Check the crankpin clamping surface and the rear crankshaft clamping surface for galls or burrs. Crankshaft clamp bolts must be carefully examined for evidence of stretching and signs of thread failure. The crankpin plug, when used, should be checked for proper setting and for cracks. Bent crankshafts must always be replaced. A final check for straightness should be made during the assembly of the engine.

The master rod, link rods, or connecting rods should be carefully examined for burrs, cracks, and for general condition. They should be

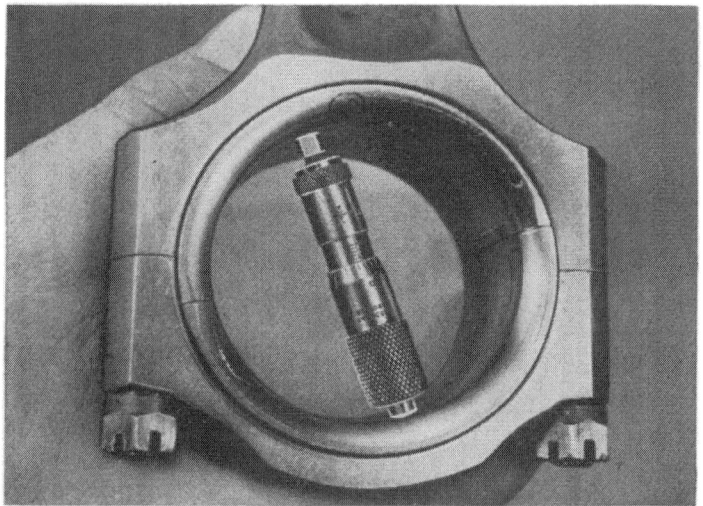

Fig. 104. Measuring the diameter of a connecting-rod bearing. (Courtesy Ranger Aircraft Engines)

checked for straightness. Knuckle-pin holes should be checked for size, nicks, and roundness. Master-rod and connecting-rod bearings should be checked for scoring, flaking, scratches, and excessive or uneven wear. Bearings which show excessive wear, scoring, and signs of flaking should be replaced. The inside of master-rod bearings and connecting-rod bearings should be measured with an inside micrometer at several locations. The thrust surface of the bearing should be examined for galling or burrs. Usually, master-rod bearings should be replaced only by the factory or an approved repair station.

The bearing plate and locking screws should be checked to make sure that they are properly secured and in good condition. Excessive misalignment of the master rod is an indication of severe damage or abuse and is cause for a replacement. No attempt should be made to straighten a master rod. The oil passages of the master rod should be open and

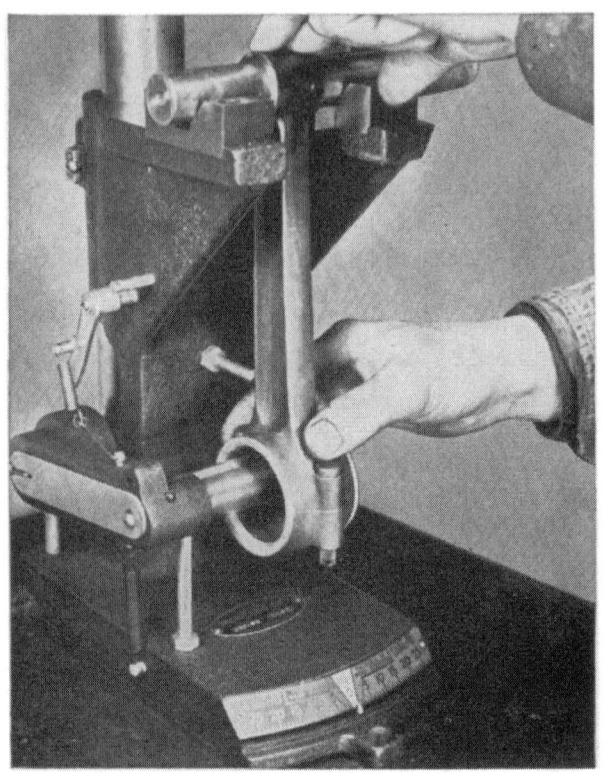

Fig. 105. Checking the alignment of a connecting rod. (Courtesy Ranger Aircraft Engines)

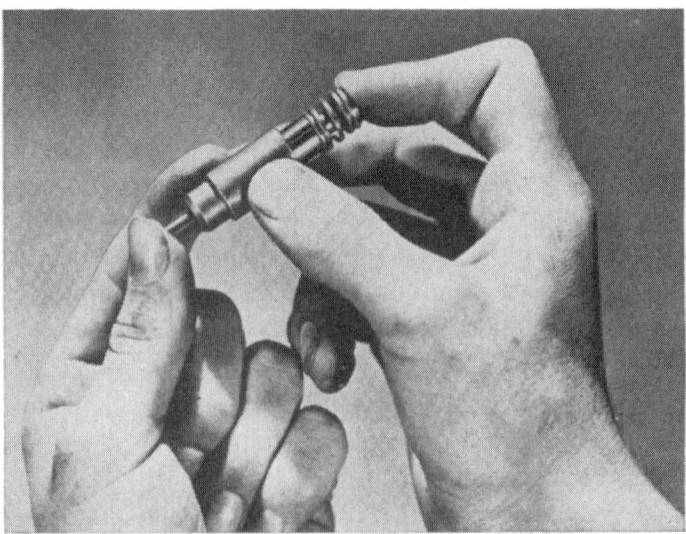

Fig. 106. Checking the cylinder movement of a hydraulic valve-lifter unit. (Courtesy Eaton Manufacturing Company)

clean. Knuckle pins should be examined for scores, burrs, and being out-of-round. The parts fitting in the master rod are checked for chafing and size. Aluminum link rods are usually replaced after approximately 1000 hr. of service, which means approximately every other major overhaul. All link rods should be checked for twist and the proper alignment of pin holes. Any distortion is cause for rejection of the rod.

The tappet guides should be examined for wear at guides and slots. The tappet guides will usually be found to be slightly oversize at the outer end of the guide. This oversize often results from hand reaming, but it will not affect the satisfactory operation of the tappet in the guide. The guides should be checked for looseness, and they should not be removed unless they are loose, cracked, worn, or damaged badly in the bore. The tappet should be inspected for scores, scratches,

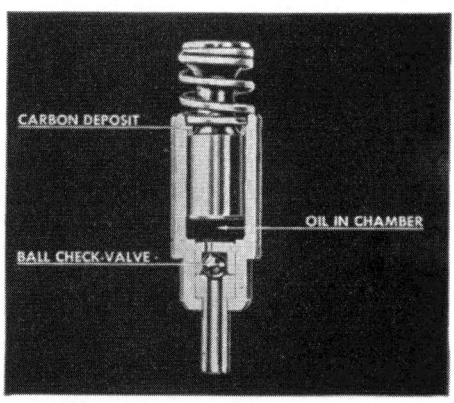

Fig. 107. Carbon deposits may prevent the removal of the cylinder of the hydraulic valve-lifter unit. (Courtesy Eaton Manufacturing Company)

cracks, and excessive wear at the ball socket. The ball socket should be checked for tightness of the ball in the tappet. The tappet assembly should be Magnafluxed for possible cracks. Tappet rollers should be checked for flat spots, wear, pitting, or cracking. The tappet pins should be checked for pitting and wear.

If the engine is equipped with a constant-speed propeller, the oil fittings should be checked to see that they are tight in the bottom of the case. If a two-position, controllable-pitch propeller valve is used, check the valve to see that it is held tightly in the front case and that it operates freely. The packing nut should be removed, and new packing installed.

All propeller oil-seal rings should be replaced at each major overhaul. If wear of the oil-seal sleeves has produced a slight ridge in the outer oil sleeve, it should be removed by the use of a fine stone. If a large ridge is present, the outer oil-seal sleeve should be replaced.

The cam track should be carefully checked for evidence of flaking, excessive pitting, wear, or cracks. If the cams show excessive wear or poor condition of the track, the entire cam assembly should be replaced. Any rivets should be checked for tightness. The cam bearing should be checked for wear or scoring.

The thrust nut, inner propeller oil seal, and thrust bearing should be checked for galling or excessive wear. The thrust nut should be carefully examined for damage to the threads or for nicks.

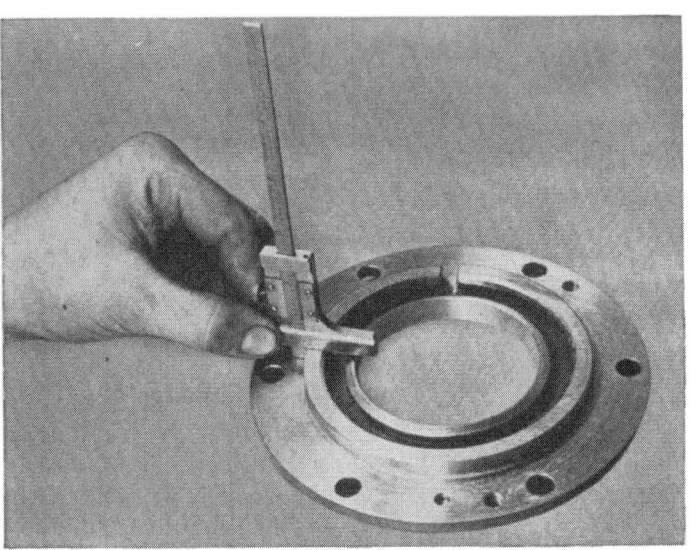

Fig. 108. Measuring the distance from the front face of the front crankcase section to the front face of the outer thrust-bearing race. (Courtesy Jacobs Aircraft Engine Company)

Any cast-in oil lines should be blown out to be sure that they are clean and free of obstructions. On most engines, the parts of the main crankcase are not interchangeable, but should be kept together in pairs. If either part needs replacing, the entire crankcase should be replaced. Cylinder pads should have any markings removed with a fine stone and be checked for flatness. The entire crankcase should be carefully checked for cracks. Particular attention should be given to the areas around the mounting lugs and cylinder pads and on top of the manifold inlet elbow. All mounting pads should be checked for smoothness, flatness, and cracks.

The entire pump assembly should be carefully inspected. The studs should be checked for tightness, bending, and damage to threads. The pump gears should be examined for marks and burrs. The shafts and keys should be examined for roughness, burrs, and wear. All parts should be carefully examined for cracks, burrs, condition of threads, and scored areas. The check-valve seats should be examined for nicks or improper seating. If the check-valve seats or balls plainly show wear, they should be replaced. The relief and check-valve springs should be examined for condition and proper tension. Any leather seals in this assembly should be discarded and replaced by new parts.

The oil-feed lines should be examined for dents and proper fit into

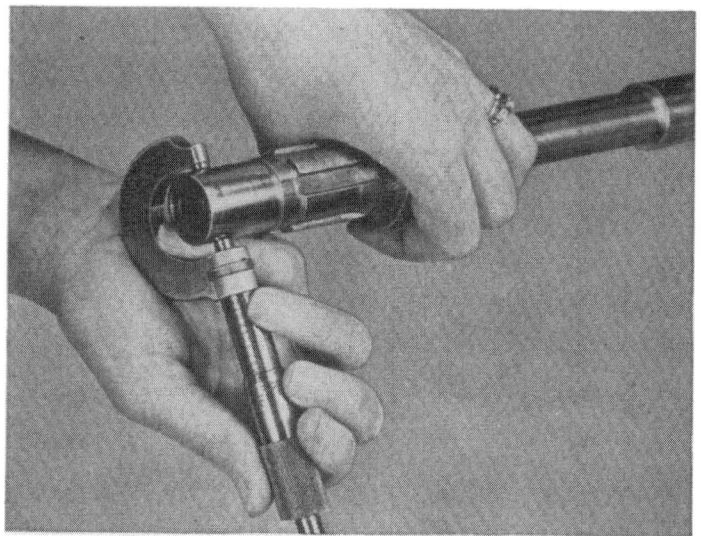

Fig. 109. The end of the shaft should be measured with a micrometer both before and after installing a cup plug. (Courtesy Ranger Aircraft Engines)

the castings. The dog on the starter gear should be examined for wear, cracks, or other damage.

The aluminum plugs in the magneto drive gear and generator drive gear should be checked for tightness by tapping lightly against the plug. Cork plugs should be replaced at each overhaul. All parts should be examined for excessive wear, pitting, or galling. Bearing surfaces of gears should be examined for scoring or excessive wear.

Unless the fittings in the rear case are left in place, the oil inlet bushings should be very carefully examined for cracks and for the condition of threads.

All mounting pads should be carefully inspected. The accessory drive is not usually disassembled. This unit may be inspected by visual examination, by rotating the various parts, and by examining for excessive wear, roughness of gears, and tightness on their shafts. If any defective parts are found, the accessory drive unit should be disassembled and the parts replaced. The rear, intermediate bearing plate should have the bushing, liner, and oil feed bearings carefully examined for scoring, wear, damage, or looseness.

Cracks are apt to occur in the ribs or on bearing bosses. All oil passages should be blown out to be sure that they are clear of obstructions.

Cylinder and piston assemblies require careful inspection. A careful check should be made to be sure that all sand from sandblasting has

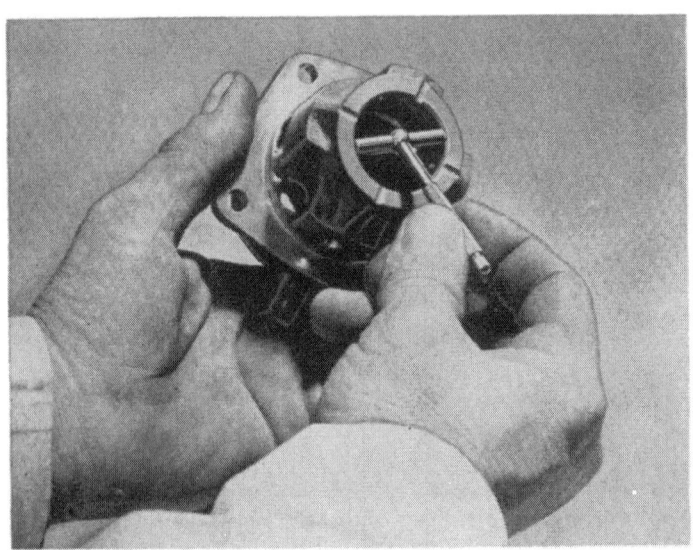

Fig. 110. Measuring the inside diameter of the vertical-drive-support bearing surfaces. (Courtesy Ranger Aircraft Engines)

been removed. The crevice between the head and end of the barrel should be examined for traces of sand. The cylinder head should be examined for cracks or nicks and for broken and cracked fins. Slightly damaged fins may be repaired. Loose spark-plug bushings should be replaced. Intake- and exhaust-valve guides should be carefully gauged, and the valve stems measured. This clearance should be checked with the table of limits. The valve seats should be examined for excessive wear, burning, and pitting, although they are not usually in such bad condition that they need replacing. The valve guides and valve seats should be replaced when excessively burned, pitted, or worn.

The cylinder barrel should be examined for damaged fins. If a cylinder-barrel fin is broken at the root of the fin, the cylinder should be replaced. The cylinder barrel should be examined for scoring and for corrosion. Badly scored cylinder barrels should not be repaired except by regrinding to oversize. This regrinding operation is usually possible but once for each cylinder barrel. Limits for wear, taper, and out-of-round are given in the table of limits. Many cylinders are constructed with choked barrels. These barrels are ground straight before assembly with the head, but are tapered at the top of the barrel by the shrinking on of the head. The regrinding of cylinders should usually be done at the factory or at an approved repair station. All moderate

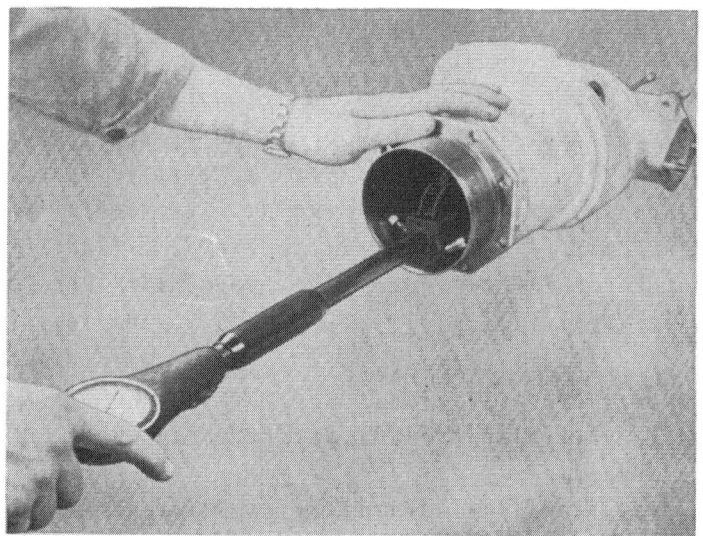

Fig. 111. Checking the bore of a cylinder by the use of a cylinder gauge. (Courtesy Ranger Aircraft Engines)

scoring or scratching can be satisfactorily repaired by stoning and the use of wet-or-dry sandpaper. Scores and scratches near the top of the piston travel are more critical than when the damage is located near the center and bottom of the barrel.

Exhaust valves are subjected to very severe heat conditions when the engine is being operated. Exhaust valves are exposed to the full heat of the burned gasses. Any considerable amount of pitting or corrosion on the stem or neck of the valve is cause for the replacement of the valve. An exhaust-valve stretch gauge should be used to check exhaust valves. The lock groove should be checked for nicks, pick-up, or wear. The tip of the valve should be examined for excessive wear or scoring. Valve seats that are badly burned or pitted should be replaced. The clearance of the exhaust valve in the guide should be determined.

It is not usually necessary to use the stretch gauge on intake valves. The clearance of the intake-valve stems in the guide should be checked.

All valve springs should be carefully examined for indications of breaks, pitting, or rust. The tension of each valve spring should be checked to be sure that it is within the limits allowed. The split-valve locks should be examined for wear, nicks, and scoring, and the valve-spring washers should be examined for damage.

The rocker-arm assembly and push rods should be carefully examined for cracks and damage. The tapered, roller, rocker-arm bearing and ball races should be examined for smoothness. A slight Brinelling may take place in the outer bearing race due to the oscillating motion of the rocker arm. A slight roughness may be allowed but, if the bearing is decidedly rough when rotated, it should be replaced. The valve adjusting screw should be examined for looseness, excessive wear, cracks, or damaged threads. The push rods should be inspected for straightness and for tightness of the balls in the end of the rod. The ball ends should be checked for cracks, roughness, or excessive wear. The rocker-arm covers should be examined for cracks around the nut recesses and in the ribs.

The intake pipes should be checked for dents and cracks. Cracks are most likely to occur at or in the flange. Cracked or badly dented pipes should be replaced.

The pistons should be inspected for cracks or scores. The piston lands, skirts, and piston-pin bosses should be checked for cracks and distorting. Cracked, badly scored pistons or pistons with broken lands should be replaced. The piston diameter should be carefully measured at various points with a micrometer. The table of limits should be consulted to determine the allowable clearance between the piston and cylinder barrel. The piston-pin bearings should be examined for scoring or wear.

The piston ring-grooves should be checked to make sure that all carbon has been removed. The groove should be examined for smoothness, flatness, and excessive wear. The groove should be carefully examined for cracks which might occur at the base of the lands at the bottom of the groove. Groove wear may be found by inserting a standard piston ring in the groove and measuring the clearance with a feeler gauge. Clearance should be measured completely around the piston. The top grooves may be remachined to use an overwidth ring by removal of material from the top of the groove. Other grooves should not be widened because this weakens the lands. Overwidth rings are not usually required. It is good practice to replace all rings at each overhaul.

If the rings are to be reinstalled, they should be carefully inspected for scores and grooves on the face, feather edges, or loss of tension. Loss of tension is indicated by a narrow free gap. Rings which are twisted should be replaced.

The piston pins should be carefully examined for correct diameter,

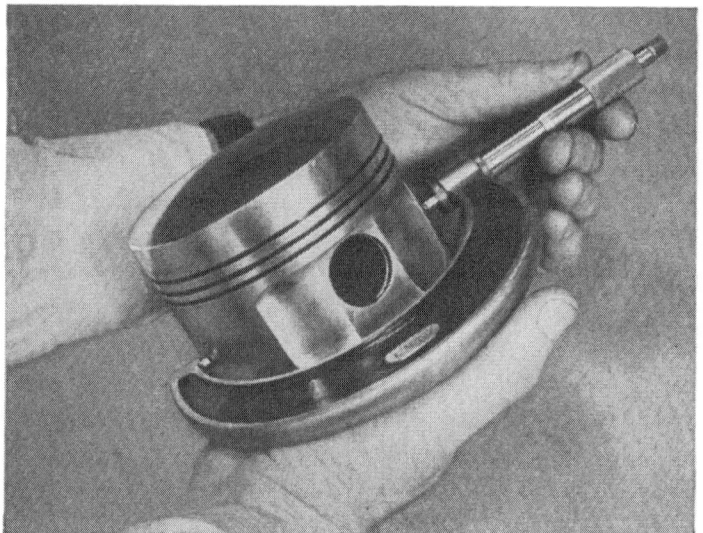

Fig. 112. Using a micrometer to measure the diameter of a piston. (Courtesy Ranger Aircraft Engines)

Fig. 113. Checking the side clearance of a piston ring. (Courtesy Jacobs Aircraft Engine Company)

cracks, scoring, wear, or out-of-round. The piston pins should also be carefully checked for straightness. If the piston-pin plug is loose in the piston pin, it should be replaced. Any damage to the piston-pin plug is cause for rejection.

The ignition manifold should be carefully examined for cracks and

Fig. 114. Checking the end clearance of piston rings by placing the rings in the proper cylinder. (Courtesy Ranger Aircraft Engines)

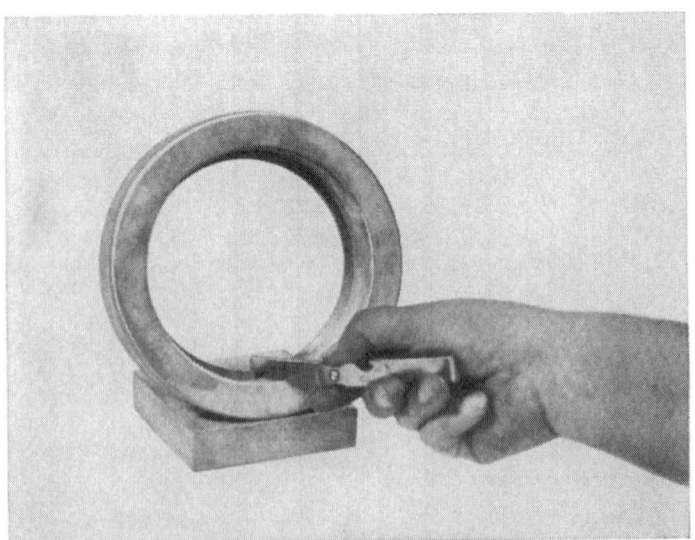

Fig. 115. Checking the end clearance of a piston ring in a piston-ring gauge. (Courtesy Jacobs Aircraft Company)

dents. Shielded conduits should be checked for wear or breaks. Spark-plug elbows and terminal sleeves should be examined for cracks and other damage. All threads should be in good condition.

All exterior lines, such as oil-sump lines, rocker scavenger and pressure lines, and primer lines should be carefully examined for breaks, dents, and corrosion. Inspect all fittings for cracks, damaged threads, and smoothness of seats for cone fittings. All rubber hose should be replaced at each overhaul.

Typical Table of Limits

The following is a typical table of limits. This table of limits is furnished by the Jacobs Aircraft Engine Company. The minimum and maximum columns contain the desired fits and clearances of the new parts. The figures in the replacement column indicate the allowable limit to which parts may wear before replacement is necessary at overhaul. An asterisk (*) under the column headed "Replace" indicates that replacement is necessary if any looseness is found at overhaul.

TABLE OF LIMITS

FIT LOCATION AND NAME	MIN. FIT	MAX. FIT	REPLACE
Piston in Cylinder (Top Land	0.0380	0.420	
(Second and Third Lands	0.0380	0.420	
(Skirt	0.026	0.029	0.038
Rings—Piston—in Grooves—All Cylinders:			
Side Clearances (Top Groove	0.0045	0.0055	
(Second Groove	0.0045	0.006	
(Third Groove	0.0045	0.006	
(Fourth Groove, Cyl. No. 1	0.005	0.0065	
(Fourth Groove, Cyls. Nos. 2–7	0.005	0.0065	
Gap—All Grooves—All Cylinders	0.070	0.075	
Pin—Piston (Piston—Hand push fit (select)	0.0000	0.005L	0.004L
(Master Rod	0.0015L	0.0022L	0.005L
(Link Rod	0.0015L	0.0022L	0.004L
Pin—Knuckle—Link Rod	0.00175L	0.0025L	0.004L
Master Rod	0.00035T	0.0010T	*
Rod—Link in Master Rod (End Clearance)	0.0170L	0.0230L	0.035L
Rod—Master to Crankpin (Diameter)	0.004L	0.0055L	0.008L
Rod—Master to Crankpin (End Clearance)	0.018L	0.024L	0.040L
Liner—Main Bearing in Crankcase	0.0040T	0.0070T	*
Bearing—Main—Roller—In Liner	0.0001T	0.0018T	0.004L
Crankshaft—to Main Bearing	0.0003L	0.008T	0.001L
Liner—Thrust Bearing to Front Case	0.0040T	0.0070T	*
Bearing—Thrust—in Liner	0.0001L	0.0015T	0.003L
Crankshaft—to Thrust Bearing	0.0005L	0.0006T	0.001L
Plate—Thrust—to Front Case (Diameter)	0.0000	0.0020L	0.008L

TABLE OF LIMITS (*Continued*)

FIT LOCATION AND NAME	MIN. FIT	MAX. FIT	REPLACE
Guide—Valve in Cylinder Head (Intake)	0.0005T	0.0015T	*
Guide—Valve in Cylinder Head (Exhaust)	0.0005T	0.0015T	*
Valve—Intake—in Guide	0.0020L	0.0035L	0.008L
Valve—Exhaust—in Guide	0.0020L	0.0045L	0.010L
Seat—Valve—in Cylinder Head (Intake)	0.0060T	0.0090T	*
Seat—Valve—in Cylinder Head (Exhaust)	0.0060T	0.0090T	*
Bearing in Rocker Arm	0.0007T	0.0017T	0.0015L
Shaft—Rocker Arm—in Bearing	0.0002L	0.0010L	0.002L
Shaft—Rocker Arm—in Cylinder Head	0.0002L	0.0010L	0.002L
Ball End—in Push Rod (Intake and Exhaust)	0.0015T	0.0035T	*
Barrel—Cylinder—in Crankcase	0.0040L	0.0070L	
Barrel—Cylinder—out-of-round and taper (Excludes cold taper at head joint)			0.006
Cam Hub—to Cam Bearing (Diameter)	0.0024L	0.0035L	0.0065L
Cam Hub to Cam Bearing (End Clearance)	0.0100L	0.0140L	0.025L
Bearing—Pinion Shaft—to Front Case	0.0006L	0.0010T	0.003L
Bearing—Pinion Shaft—to Front Case (Large)	0.0006L	0.0010T	0.003L
Shaft—Cam Pinion—to Bearing	0.0001L	0.0007T	0.001L
Gears—Timing and Cam Drive (Backlash)	0.0040	0.0080	0.020
Gears—Cam Pinion and Cam Ring (Backlash)	0.0040	0.0080	0.020
Guide—Tappet—in Front Case (Intake and Exhaust)	0.0000	0.0010T	0.001L
Tappet in Guide	0.0003L	0.0009L	0.003L
Pin—Tappet Roller—In roller	0.0013L	0.0020L	0.005L
Pin—Tappet Roller—In tappet	0.0013L	0.0020L	0.005L
Gears—Oil Pump (End Clearance)	0.0025L	0.0040L	0.008L
Gears—Oil Pump (Backlash)	0.0040	0.0100	0.020
Gears—Oil Pump—on Idler Shaft	0.0005L	0.0015L	0.003L
Oil Pump—Pressure Section—to Rear Case	0.0020L	0.0040L	
Bushing—to Oil-Pump Pressure Section	0.0010T	0.0030T	*
Shaft—Oil-Pump Drive—to Bushing and (Press. Sect.) Pump	0.0015L	0.0030L	0.005L
Shaft—Oil Pump Drive—to Bushing and (Scav. Sect.) Pump	0.0015L	0.0030L	0.008L
Bushing—to Oil-Pump Scavenger Section	0.0003L	0.0012L	0.003L
Shaft—Idler—to Oil Pump (End Section)	0.0005L	0.0010T	0.002L
Shaft—Idler—to Oil Pump (Center Section)	0.0000	0.0015L	0.0025L
Shaft—Drive—to Oil Pump (Center Section)	0.0010L	0.0025L	0.004L
Front Case to Crankcase	0.0000	0.0040T	0.008L
Crankcase—Front Half to Crankcase Rear Half	0.0000	0.0040T	0.002L
Rear Case to Crankcase	0.0010L	0.0040L	
Plug— (Small) in Crankshaft (Front and Rear)	0.0010T	0.0035T	*
Plug— (Large) in Crankshaft (Front)	0.0010T	0.0040T	*
For Replacement, Fit Plug to	0.0010T	0.0025T	*
Gear—Starter to Bushing (End Clearance)	0.0060L	0.0260L	
Gear—Generator Drive to Bushing (End Clearance)	0.0060L	0.0300L	
Gear—Magneto Driven to Bushing (End Clearance)	0.0370L	0.0750L	
Gear—Oil-Pump Drive to Bushing (End Clearance)	0.0350L	0.0690L	

TABLE OF LIMITS (Continued)

FIT LOCATION AND NAME	MIN. FIT	MAX. FIT	REPLACE
Gears—Crankshaft Cluster and Oil Pump Drive (Backlash)	0.0040	0.0080	0.018
Crankshaft to Oil-Feed Bearing	0.0015L	0.0030L	0.0045L
Bearing—Oil Feed to Rear Int. Bearing Plate	0.0010T	0.0030T	*
Bushing—Oil-Pump Drive Gear to Rear Intermediate Bearing Plate	0.0010T	0.0030T	*
Gear—Oil-Pump Drive in Bushing	0.0010L	0.0020L	0.004L
Gear—Generator Drive in Bushing	0.0020L	0.0030L	0.005L
Gear—Starter—in Bushing	0.0010L	0.0020L	0.004L
Bushing to Master Rod (Piston Pin End)	0.0010T	0.0020T	*
Bushing to Link Rod (Piston Pin End)	0.0010T	0.0020T	*
Plug to Piston Pin	0.0010T	0.0020T	*
Plug to Piston Pin (Service Replacement) Fit to	0.0000	0.0010T	*
Bearing to Master Rod	0.0010T	0.0020T	*
Plate—Thrust to Front Case (Peel Shim to Obtain Clamp Fit of)	0.0020T	0.0040T	*
Bushing—Gen. Drive and Magneto Driven Gear to Rear Intermediate Bearing Plate	0.0010T	0.0030T	*
Gear—Magneto Driven in Bushing	0.0020L	0.0030L	0.005L
Seal—Oil—to Rear Case (Mag. and Gen. Drive)	0.0010T	0.0070T	*
Plug—Gen. and Mag. Drive Gears (Alum. Plug)	0.0020T	0.0045T	*
(Cork)			*
Gear—Timing to Crankshaft	0.0004L	0.0014L	0.002L
Bearing—Cam to Crankshaft	0.0004L	0.0019L	0.0025L
Gear—Crankshaft Cluster to Crankshaft	0.0000	0.0010L	0.0016L
Gear—Magneto Drive to Crankshaft	0.0000	0.0010L	0.0016L
Gears—Crankshaft Cluster & Gen. (Backlash)	0.0040	0.0080	0.018
Liner—Rear Intermediate Bearing Plate	0.0030T	0.0060T	*
Bearing—Ball in Liner—Rear Int. Bearing Plate	0.0010L	0.0008T	0.004L
Plate—Rear Int. Bearing to Crankcase	0.0000	0.0030T	0.002L
Gears—Magneto Drive & Mag. Driven (Backlash)	0.0040	0.0080	0.018
Gear—Crankshaft Cluster & Starter (Backlash)	0.0040	0.0080	0.018
Bushing—Starter Gear to Rear Int. Bearing Plate	0.001T	0.004T	*
Bearing—Rear Ball to Crankshaft	0.0005L	0.0005T	0.001L
Gears—Oil Pump to Pump Bodies (Diametral Clearance)	0.003L	0.005L	0.010L
Body—Propeller Valve to Front Case	0.0005L	0.0015L	0.002L
Valve—Propeller to Body	0.0002L	0.0009L	0.002L
Shaft—Propeller Valve to Body	0.0005L	0.0015L	0.0025L
Fitting—Propeller Oil to Front Case	0.0005L	0.0015L	0.002L
Sleeve—Inner to Crankshaft	0.0004L	0.0019L	0.0025L
Sleeve—Outer to Front Case (Shrink Fit)	0.000T	0.0015T	*
Rings—Oil Seal (Side Clearance, Total Four Rings in Groove)	0.0055L	0.0115L	
Rings—Oil Seal (Cast Iron) —Gap	0.008	0.016	
Rings—Oil Seal (Bronze) —Gap	0.006	0.014	
Bushing—to Link Rod (Knuckle-Pin End)	0.0015T	0.0025T	*
Bearing—Gen. Drive Gear in Rear Case	0.0001L	0.001L	0.003L

TABLE OF LIMITS (Continued)

FIT LOCATION AND NAME	MIN. FIT	MAX. FIT	REPLACE
Gear—Generator Drive to Bearing	0.0001L	0.007T	0.001L
Key—Oil-Pump Drive Shaft			
Side Clearance in Shaft Fit to	0.0005T	0.0015L	0.003L
Side Clearance in Gears Fit to	0.0015L	0.0040L	0.006L
Radial Clearance in Gears	0.0005L	0.0143L	
Key—Cluster Gear			
Side Clearance in Crankshaft Fit to	0.0000	0.0010L	0.002L
Side Clearance in Magneto Drive and Cluster			
Gears Fit to	0.0000	0.0020L	0.003L
Radial Clearance in Gears	0.004L	0.0140L	
Key—Timing Gear			
Side Clearance in Crankshaft Fit to	0.0000	0.0020L	0.003L
Side Clearance in Gear Fit to	0.0000	0.0020L	0.003L
Radial Clearance in Gear	0.0035L	0.0160L	
Key—Cam-Drive Gear			
Side Clearance in Pinion Shaft Fit to	0.0000	0.0010L	0.003L
Side Clearance in Cam-Drive Gear Fit to	0.0000	0.0030L	0.004L
Radial Clearance in Gear	0.0030L	0.0170L	
Tube Oil in Rear Intermediate Bearing Plate	0.0020T	0.0035T	*
Tube Oil in Rear Case	0.0010L	0.0020L	
Maximum Run-Out, Rear Cone Seat		0.005	
Maximum Run-Out, Front Cone Seat		0.010	

SPRING SPECIFICATIONS

DESCRIPTION	WIRE Dia.	MINIMUM LOAD WHEN NEW		MIN. LOAD BEFORE REPLACEMENT
		Valve Closed	Valve Open	Valve Open
Spring—Valve Outer	0.170	58# at 1–7/8″	88# at 1–13/22″	85# at 1–13/32″
Spring—Valve Inner	0.125	33# at 1–7/8″	48# at 1–13/32″	46# at 1–13/32″
Spring—Relief Valve	0.0625	8–1/2#–9# at 2–1/8″		
Spring—Check Valve	0.034	1/2# at 1–13/16″		
Spring—Check Valve	0.031	.19# at 1–7/8″		

TORQUE LIMITS, ETC.

Crankshaft Clamp Bolt, Part No. 30100 Stretch to	0.012″ min. or next cotter hole
Crankcase Bolt Nuts	325–350 in. lb.
Cylinder Hold-Down Nuts	250–300 in. lb.
Rocker-Shaft Nuts	120–130 in. lb.
Rocker-Cover Stud Nuts	40– 50 in. lb.
Spark Plugs	300–360 in. lb.
Thrust Nut	550–650 ft. lb.
Nut—5/8″—18, Special (Cam-Drive Pinion Shaft)	400–600 in. lb.
Valve-Clearance Adjusting-Screw Lock Nut	300–325 in. lb.

XI | MAGNETIC INSPECTION

Magnetic inspection is the method of inspection commonly called Magnaflux inspection. This method of inspection is used on parts made of ferrous metals, that is, iron and its alloys. This method depends upon the magnetic effects of an electric current. Parts made of iron alloys take on the characteristics of a magnet when placed in a magnetic field. The magnetic lines of force flowing through the metal part, when interrupted by a crack or break in the part, cause the edges of the crack to become little individual magnets. Each edge of the crack becomes a magnetic pole. The interruption of the flow of the magnetic lines is not visible to the naked eye, but when a magnetic substance, such as fine iron powder or powdered iron oxide, is brought into contact with the part, the powder will cling to the edges of the crack, making it visible. A fine invisible crack will then appear as a black line in the surface of the metal.

Several methods are used in Magnaflux inspection of various engine parts. The most commonly used method is based upon the circular magnetic field which develops about any conductor carrying an electric current. The part to be magnetized is mounted between two contact heads, and a low-voltage, high-amperage current is passed through the part to be inspected. This will cause a circular magnetic field to be set up around the outside of the part. This circular field will tend to show up cracks running parallel to the long axis of the part. If the part is so shaped that it cannot be readily mounted between the contact heads, it may be suspended on a copper rod between the heads. This method of mounting is used for such parts as a ring-shaped bearing race. A circular magnetic field will be induced around the copper rod and the same magnetic field will be induced in the ring mounted on it.

Another method of magnetizing is brought about by placing the part to be inspected inside a coil through which the low-voltage, high-amperage current is passed. The magnetic lines of force will flow through the center of the coil, and any object within the coil will be magnetized. The part will be magnetized in a direction which is at right angles to the coil. Cracks show up best when they run at right

Fig. 116. An engine valve receiving the Magnaflux treatment. (Courtesy Ranger Aircraft Engines)

Fig. 117. Magnafluxing a circular engine part. (Courtesy Ranger Aircraft Engines)

angles to the direction of the magnetic lines within the part. A crack which runs at less than 30° to the lines of force will not be shown clearly. A part should be tested by turning it in several directions within the magnetic field to detect cracks which run at various directions in the part. It is important that the part be demagnetized before it is tested in a new direction. Parts may be magnetized either by a direct current or an alternating current. An alternating current is necessary to demagnetize the part.

There are a number of different types of Magnaflux machines. The most commonly used type for the inspection of aircraft engine parts is the horizontal type, which has a bench, below which are mounted the various controls and automatic timing devices and a tank and pump for the inspection fluid. On top of this bench are mounted two contact heads, one of which is fixed and the other movable along a track on top of the bench. A movable coil is also mounted on the bench. This coil can be moved along the central track and fits over the fixed head when not in use. A foot switch is usually provided to turn the current on and off.

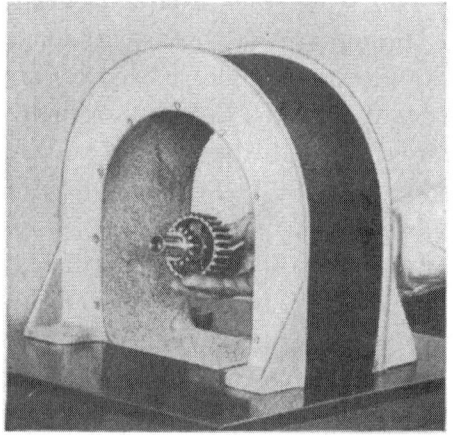

Fig. 118. An engine part in the magnetic field of the Magnaflux equipment coil. (Courtesy Ranger Aircraft Engines)

In order to magnetize a shaft, one end is placed against the stationary head, and the other end touches the movable head, which then holds it in place. As the switch is closed to turn on the current, the plate in the stationary head is automatically forced against that end of the part by compressed air. This action makes sure that there is good contact between the heads and the part. Small parts which need to be magnetized may be held within the coil by hand. The part should be held close to the inner edge of the coil where the magnetic lines of force are strongest. Parts which are heavy or difficult to hold can be clamped between the heads or mounted on a rod between them.

The inspection material used is a finely powdered black or red iron oxide. The fluid is a noninflammable liquid such as safety naphtha. When a–c equipment is used, the same coil may be used to demagnetize the parts. If direct current is used, a special a–c demagnetizing coil must be provided.

To demagnetize a part, the alternating current in the coil should be turned on and the part moved from a point about 18 in. in front of the coil through the coil to a point about 18 in. back of the coil. The part should be moved slowly and evenly through the coil at a speed of about 12 ft. per min. for large parts. Small parts may be moved through the coil at about twice this speed. It is important that all parts be demagnetized before being installed in the engine. If they are not thoroughly demagnetized, they may attract small steel particles which may be present in the engine and damage the bearings.

To test whether a part has been thoroughly demagnetized, it should be brought close to a small magnetic compass. The compass needle should not be deflected more than 3° at a distance of 6 in. from the part. If the part is not thoroughly demagnetized, it should again be passed through the coil. In passing the part through the coil a second time, it should be held in a different position from that in which it was held the first time.

Only a skilled operator can be relied upon to interpret Magnaflux indications. There are no definite rules which may be followed, and satisfactory results depend upon the skill of the inspector. Magnafluxing should be performed only by a person thoroughly trained in this work. All Magnaflux indications should be examined very carefully. In many cases, close inspection will show that some of the indications may be ignored. Forging laps and seams in the metal may cause Magnaflux indications which are in themselves of no importance. When indications of this kind appear on highly stressed parts, they may give rise to fatigue cracks. If the current used is higher than is necessary, Magnaflux indications may appear upon the part which are not indications of cracks or failures.

A table is usually furnished which gives the correct current to be used in magnetizing each part of the engine. Almost any part will show Magnaflux indications if a high enough current is used. Some indications may be caused by the shape of the part itself. A sudden thinning in the cross section of a part may cause a Magnaflux indication. An example of this kind is a keyway. A keyway often produces Magnaflux indications which look as though there were two cracks in the material.

It is important to have good light when Magnafluxing. A high-powered magnifying glass or binocular microscope is also necessary for a thorough inspection.

The fine iron powder suspended in the liquid is injurious to the skin, and the operator's hands should be protected by a chemical compound recommended by the manufacturer of Magnaflux equipment. Rubber gloves are not recommended.

When making a Magnaflux inspection, all parts should be completely disassembled. Gears should be removed from shafts, and all parts, such as dynamic dampers, should be removed.

Antifriction bearings should never be Magnafluxed. It is impossible to remove all of the iron powder from bearings of this kind after inspection. Rocker arms containing needle bearings or roller bearings should not be Magnafluxed. If the part contains oil holes or passages, these should be sealed with heavy grease, paraffin, corks, or wooden plugs.

All parts to be Magnafluxed must be perfectly clean. Any smooth-surface cracks or scores in polished surfaces should be polished out before Magnafluxing. These tiny scratches will give the same indications as a small crack.

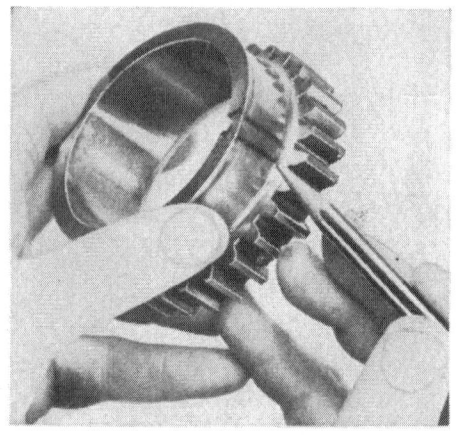

Fig. 119. A keyway cut into an engine part is indicated by the Magnaflux treatment. This is not a defect. (Courtesy Ranger Aircraft Engines)

When a number of parts are being Magnafluxed on a rod at the same time, they should be separated so that they do not touch each other. If these parts are in contact, they may be magnetized unevenly. If parts are to be Magnafluxed with the coil while suspended on a rod they must be placed firmly together. These parts become strongly magnetized and will be drawn together with great force as soon as the current starts to flow.

A slight indication of failure may be found from which it is impossible to determine whether or not there is actual damage present. It is sometimes possible to stone out these indications when they are only a few thousandths of an inch deep. Stoning should only be done when it does not materially affect the strength of the part. A record may be preserved of slight indications by allowing the Magnaflux pattern to dry thoroughly and, after the indication is dry, pressing a strip of transparent Scotch tape down evenly over the indication. When the tape is removed, the iron particles forming the pattern will cling to it. This tape with the pattern on it is then pasted on the Magnaflux inspection report. A later Magnaflux inspection should indicate whether the crack is increasing in length or whether the damage was so slight that it may be ignored.

In Magnafluxing the parts of an engine, the manufacturer's recommendations for the current to be used on various parts should be carefully followed.

XII REPAIR AND REPLACEMENT

In this section an attempt is made to describe more or less in detail the common necessary repair and replacement operations. Whenever an engine undergoes a complete overhaul, the aim should be to restore the engine as nearly as possible to its condition when new. It is not possible in a text of this kind to explain every procedure in detail or just how to make every repair. The ability to make repairs and the judgment as to what repairs are to be made and the manner in which they are made come with much practical experience and training. The material and methods given are not intended to be all-inclusive and to fit every type of repair which is necessary. In general, however, the following may be applied to any engine and should assist the mechanic materially in his maintenance and repair work.

There are certain precautions which apply to repair work on all engines. In drilling and reaming holes, care should be taken to remove all sharp edges remaining around the hole. All burrs should be removed with a fine stone or crocus cloth. Emery cloth should never be used for cleaning or polishing engine parts. Particles of emery may be embedded in the surface of the material and continue to cause scoring and excessive wear. Emery cloth also may leave scratches on the material which will later develop into cracks leading to the failure of the engine part. Crocus cloth of a good grade or soft-backed wet-or-dry crocus paper, number 400-A, may be used for cleaning and polishing engine parts. A fine soft stone may be used for removing burrs, pits, or nicks. Nicks may often be removed by rounding out so that there is no sharp change in the direction of the surface of the material. All sharp edges must be removed. A sharp edge, particularly at the bottom of a nick, will almost certainly result in a crack when subjected to the vibration of the engine.

Bronze bearings that are in good condition should not be disturbed. Any scratches in the surface of bronze bearings should be carefully smoothed with a burnishing tool. Slightly scored bearings may be smoothed with crocus cloth. Wet-or-dry crocus paper, a fine stone, or crocus cloth may be used to repair minor damage to steel bearing surfaces and shafts. Scratched master-rod bearings should be replaced.

When bushings are installed, the alignment of the oil hole with the oil passage in the mating part can be obtained by drawing a crayon or

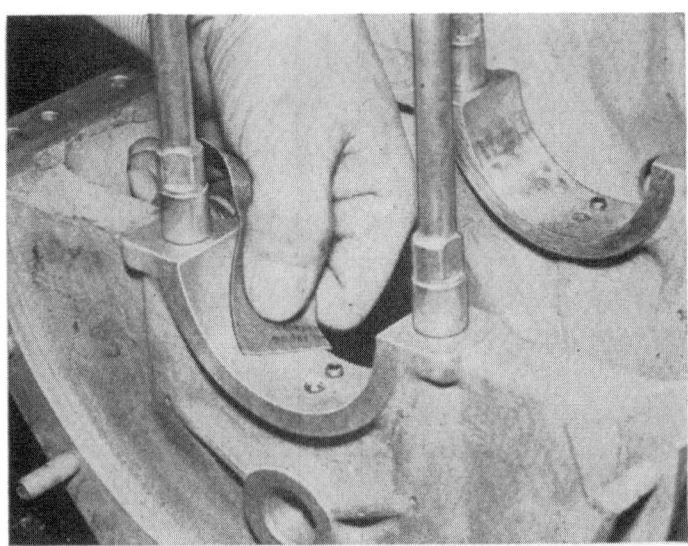

Fig. 120. Polishing out slight defects in a bearing by using crocus cloth. (Courtesy Ranger Aircraft Engines)

pencil line lengthwise on the outside diameter of the bushing which lines up with the center of the oil hole and is square with the surface of the bushing. Mark the position of the oil passage on the edge of the hole which is to receive the bushing. Line up the two marks and press the bushing into place. If paraffin is used to plug the oil holes, it should not be allowed to run into the hole in a melted condition. Melted paraffin which gets into oil holes or oil passages is very difficult to remove. Paraffin may be rolled into a small ball between the fingers and applied to the hole like putty. If this is done, no difficulty in removing all the paraffin will be encountered.

Studs which have become loose, broken, or have damaged threads, should be replaced by an oversize stud. A stud which has already been replaced by oversize should be replaced by the next oversize or by a step stud. It is nearly always possible to install the oversize stud without retapping. If a step stud is to be installed, the hole should be drilled and retapped. Stud holes which have stripped threads must be retapped for installation of step studs. It is necessary, when driving studs, to be sure that the alignment is correct and the projection proper.

Overheating should be carefully avoided when parts are heated to assist in the removal or insertion of bushings or bearings. Uneven heating may result in warping or distortion of the part. If it is necessary to heat a part, the heating should be done in a proper furnace or in an oil bath. The heating of parts with a torch is prohibited. Magnesium parts should not be heated to more than 250° F. A boiling water bath will usually be satisfactory for heating magnesium parts. As a rule, aluminum parts should not be heated over 300° F. unless the manufacturer specifically indicates

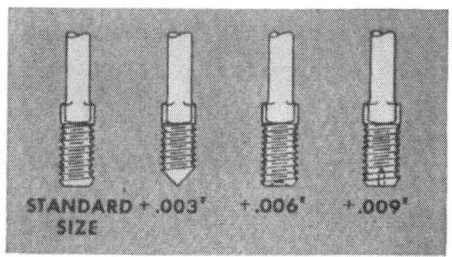

Fig. 121. Standard and oversize studs may be identified by the shape of the end which enters the casting. (Courtesy Ranger Aircraft Engines)

otherwise. If a part which contains more than one bushing, liner, or insert has been heated for the insertion of a part, all other inserts in the part should be checked to make sure that they have not been loosened by the heat. Slow cooling is necessary for all heated parts.

If an elastic stop nut is slightly loose, it may have the self-locking feature renewed by placing the nut on a smooth, flat, metal surface and striking the nut with a hammer. This operation can be performed but once with each nut and the manufacturer's recommendations should be followed.

A mixture of white lead, oil, and graphite should be applied to all pipe-thread fittings and all pipe plugs before they are installed. This mixture is made of 1 part powdered graphite mixed with 4 parts white lead. To this mixture is added enough S.A.E. No. 30 oil to form the proper paste.

Surface Treatment of Magnesium Alloys. All parts made of magnesium are given a dichromate pickle treatment by the manufacturer to protect the part from corrosion. Any magnesium part which shows signs of corrosion at overhaul should be repickled. To prepare the pickling solution, $1\frac{1}{2}$ lb. of sodium dichromate should be dissolved in $6\frac{1}{2}$ pt. of water. Stir this solution constantly while adding to it $1\frac{1}{2}$ pt. of nitric acid having a specific gravity of 1.42. The solution is used at room temperature, and the part being pickled should also be at room temperature. The part, which must be entirely free from all grease and paint, is placed in the solution for 10 to 15 sec. It is then removed, rinsed in cool running water, and washed in hot water immediately.

A container for the pickle solution should be of commercially pure aluminum, glass, or earthenware. If the part is too large to be dipped into the solution, the solution may be applied to the part with a brush or swab.

Care should be taken to wash away all excess solution thoroughly and to make sure that no solution is trapped in bends, passages, recesses,

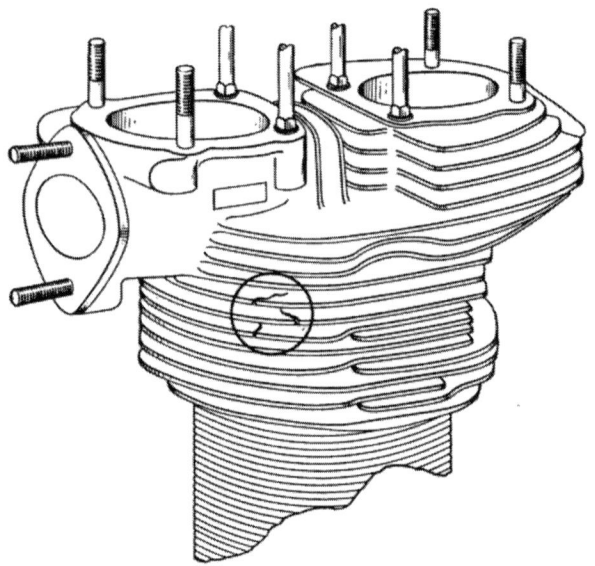

Fig. 122. A cylinder head showing small cracks across several fins. The cracked portions should be cut off and the metal smoothly finished. (Courtesy Ranger Aircraft Engines)

or corners of the part. The best method of washing a part which has been treated in this manner is to scrub it with a brush while flushing with large amounts of warm water. The manufacturer of magnesium parts may be asked for additional information, if it is required.

Painting. It is often possible to repaint a part without completely removing the old paint which may be in good condition. If the old paint is not to be removed, it should be smoothed with a fine grade of sandpaper to provide a good base for the new paint. Loose or brittle paint should always be removed. To remove all paint completely, the part should be dipped in a good paint remover. The manufacturer's directions for the use of these removers, or strippers, should be carefully followed. Cylinder heads and barrels should have all old paint removed by sandblasting and should be repainted at each major overhaul. A paint remover may be used if sandblasting equipment is not available. All paint should be applied by carefully following the manufacturer's directions.

Assembly Precautions. To guarantee continued successful operation of the engine, it is absolutely essential that the mechanic and inspector see that proper attention is given to every detail of inspection, repair, and assembly. The inspector and mechanic should continually keep in mind that the slightest neglect on their part may result in the failure of the engine in service.

Cotter pins, gaskets, leather seals, seal lines, rubber hose, and safety wire should never be used a second time. Any safety device which has been bent or worn, or shows damage of any kind should be replaced. It is essential that the greatest care be taken to prevent dirt, dust, nuts, washers, cotter pins, or any other small article from falling into the engine during assembly. Any such foreign matter may cause complete engine failure. All parts must be carefully cleaned before assembly. The use of compressed air is strongly recommended for cleaning because rags and waste leave lint and bits of thread which may clog oil lines and strainers.

Every step in the assembly of an engine must be completed before passing on to new work. Never leave a loose nut, or a cotter pin to be installed later. They are too easily overlooked. Never slack off on a nut to line up castellations with the cotter pin hole in the bolt or stud. If the nut cannot be tightened without exceeding the maximum torque, a new nut or washer must be used. It is sometimes possible to dress down the nut or washer slightly if no other thickness is available.

Cotter pins should fit tightly into the bolt or stud holes. They should not be so tight that they have to be driven into place. Cotter pins that are installed in such parts as the cam-driven pinion, crankshaft clamp bolt, crankpin bolt, knuckle-pin bolts, or other parts within the engine which are not easily accessible must be bent over tightly and in the correct manner. Cotter pins which are installed on a knuckle-pin bolt or castle nuts should have the end of the cotter pin bent over and inserted in the next castellation.

When using safety wire, a size should be selected which will snugly fit the hole through which it is placed. The wire should be twisted uniformly with pliers to obtain tight loops at each end but not tight enough to cause excessive strain. It should always be twisted tightly enough to prevent vibration which would cause breakage of the wire in service.

In all places where oil might seep out from between machined or lapped surfaces, such as between the crankcase section and parting

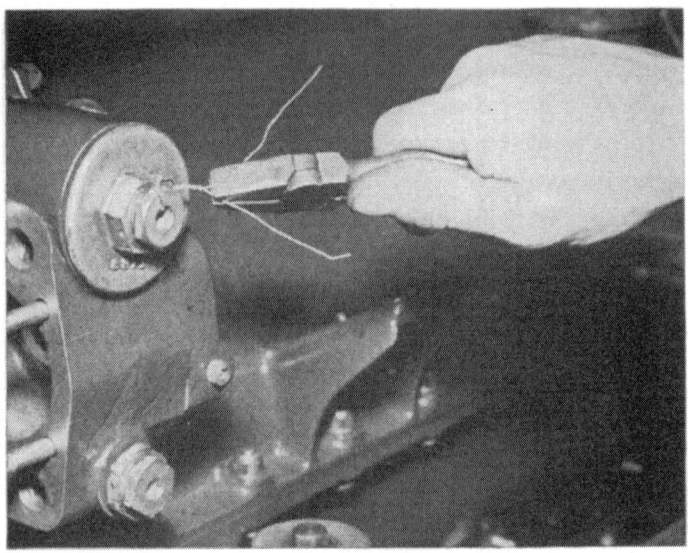

Fig. 123. The oil-strainer cover plug is safetied with safety wire. (Courtesy Ranger Aircraft Engines)

flanges, the mating surface should be coated with a good gasket paste. The gasket paste used should be of a type recommended by the manufacturer.

The rubber oil-seal rings under the cylinder hold-down flanges and the oil seals at the end of the push-rod tubes should always be replaced with new rings when removed.

The oil pump of the engine will not begin to furnish a supply of oil until the engine has been turned over a number of revolutions. To prevent the scoring of dry surfaces before the normal lubrication takes place, all surfaces normally lubricated by oil from the pump should be thoroughly covered with a good supply of engine oil at the time the parts are assembled. All parts which are a drive or push fit should be coated with oil to assist in their assembly in the engine.

Ignition Manifold. Replace any clamps, terminal sleeves, grommets, or flexible conduits which are damaged. Before reassembling the ignition manifold, new wires should be installed. The ignition cables should

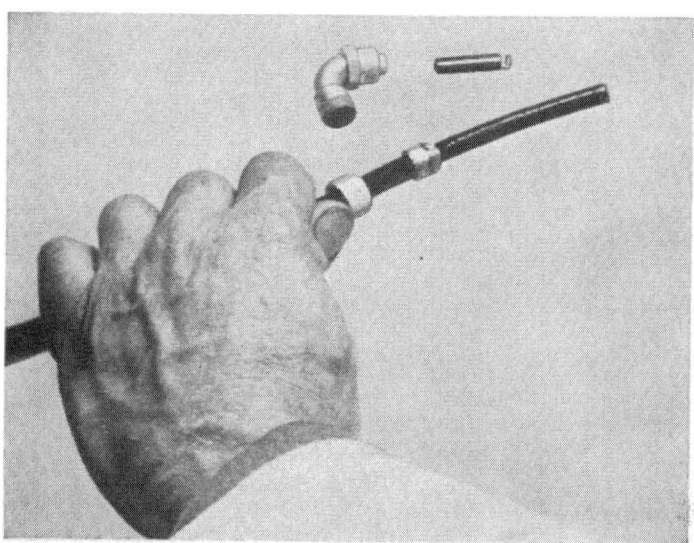

Fig. 124. The first step in replacing a spark-plug elbow on an ignition wire is to place the nut and then the cone on the wire, then fold the shielding back over the cone. (Courtesy Ranger Aircraft Engines)

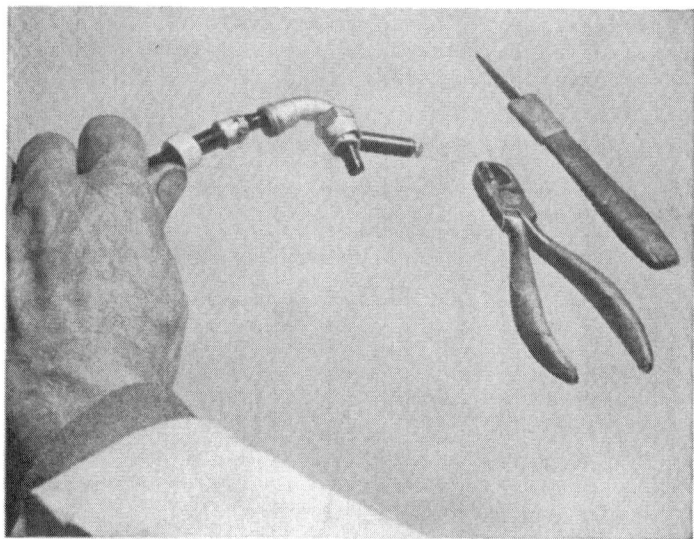

Fig. 125. The second step in replacing a spark-plug elbow is to insert the wire in the spark-plug elbow, avoiding damage to the inner insulation, then tighten the nut on the elbow. (Courtesy Ranger Aircraft Engines)

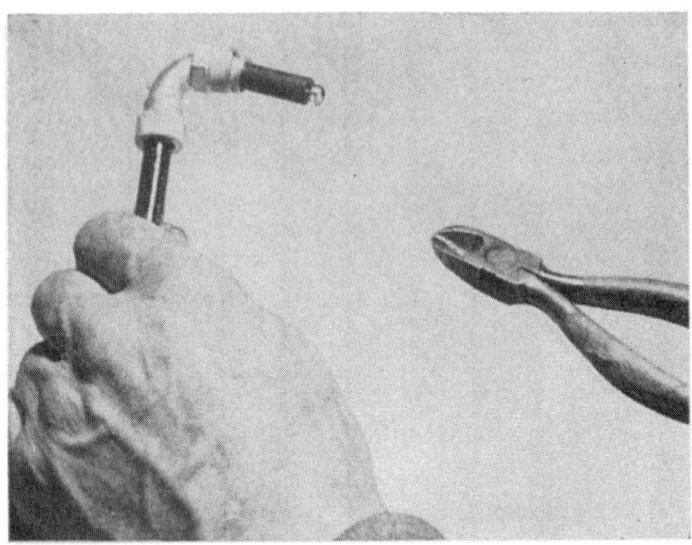

Fig. 126. The third step in replacing a spark-plug elbow is to slip the spark-plug insulating sleeve over the short end of the wire extending through the elbows and bend the wire over, then bend over the ends of the strands but do not solder. (Courtesy Ranger Aircraft Engines)

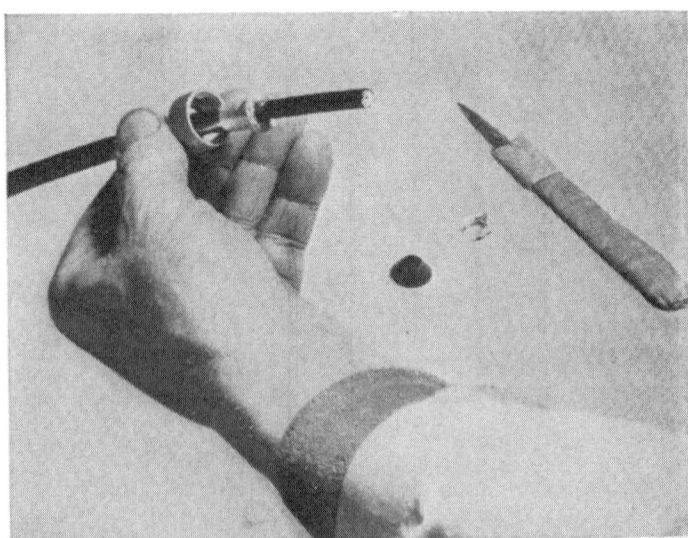

Fig. 127. The first step in installing the magneto terminal. (Courtesy Ranger Aircraft Engines)

comply with the manufacturer's recommendations as to installation and size. Care must be taken to avoid damaging the outside protective covering or insulation wire while drawing ignition cables into the manifold. All wiring should be installed in accordance with the manufacturer's manual. A careful recheck should always be made after the installation of wiring to be sure that all wires are in their proper place.

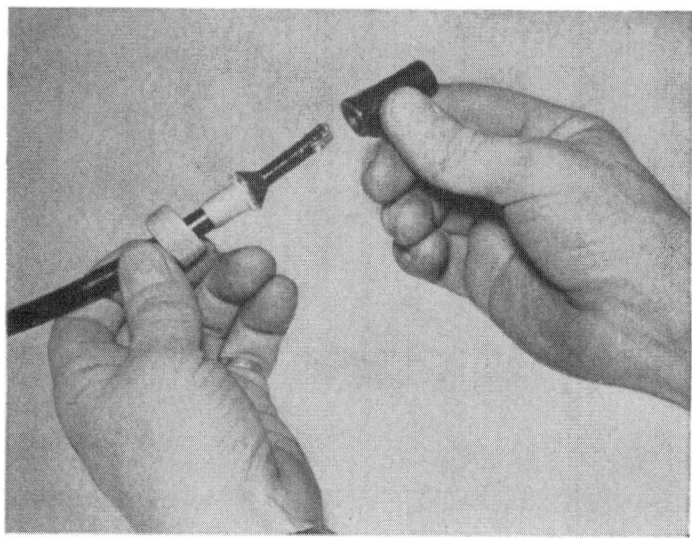

Fig. 128. The second step in installing the magneto terminal. The wires should not be soldered to the magneto terminal clips. (Courtesy Ranger Aircraft Engines)

Oil Pump. The finished surfaces of the oil-pump sections should be carefully cleaned with crocus cloth and unleaded gasoline. The tachometer bushings or leather packings should be replaced with new parts whenever they are removed. The studs should be carefully examined, and all damaged studs should be replaced. Small nicks and burrs should be removed from gear teeth with a soft stone, and the bearing surfaces of all shafts should be carefully cleaned. The keys and keyways should be carefully examined, and all burrs and nicks should be removed with a stone. If the keys are to be replaced, it is usually necessary to hand-fit the key to comply with the limits given in the table of limits.

When it is necessary to replace an oil-pump section, the mating surfaces should be checked for close fit by using Prussian Blue. The spare parts of oil-pump sections have usually already been hand-lapped on a surface plate. If any lapping is necessary to obtain a perfect fit, the old part should be lapped rather than the new. All lapping compound must be thoroughly washed off before installing the parts. Particular attention must be paid to such parts as corners, gear pockets, and shaft sockets. When assembling a pump, the mating surfaces should be treated with a sealing compound recommended by the manufacturer. Usually this sealer is only applied to a narrow strip, $\frac{1}{8}$ in. to $\frac{1}{4}$ in. wide, around

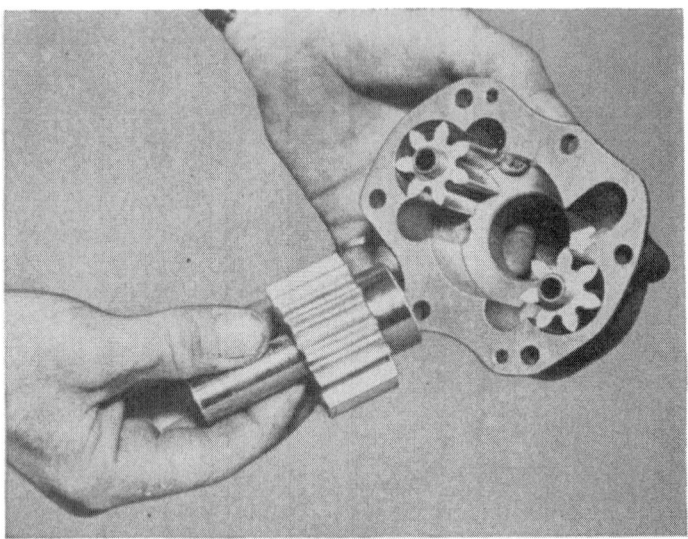

Fig. 129. Installing the scavenge-oil-pump gears. (Courtesy Ranger Aircraft Engines)

the outer edge of the mating surface. In assembling an oil pump, all nuts or screws should be tightened evenly to prevent distorting the parts. After assembly, the pump should be carefully checked to see that it does not bind at any point when the drive shaft is rotated by hand.

Rear Intermediate Bearing Plate. Stud replacements are made wherever necessary. It is important to be sure that all oil passages are clean. Bushings or liners that are slightly scratched or scored should be smoothed.

When the rear crankshaft ball-bearing liner requires replacement, the two tapered pins should be driven out. The bearing liner should be turned down to a thin shell, and this thin shell should be collapsed and removed. Any burrs or roughness should be removed from the bore of the casting. The bore in the bearing plate must be carefully measured to see whether or not it will fit a standard liner, as given in the table of limits. If the bore in the bearing plate has worn oversize, an oversize liner should be installed. The bearing plate should be heated to about

200° F., and the liner installed with a press fit. The liner bore should be bored to a fine finish and to dimensions determined from the table of allowances. All burrs and rough spots should be carefully removed.

When the whole transfer bearing requires replacement, its fastening should be removed and the bearing pressed out. The same general direc-

Fig. 130. An accessory drive-gear train. (Courtesy Ranger Aircraft Engines)

tions should be followed for installing a new bearing as those given for the rear crankshaft ball-bearing liner.

Various drive bushings in the rear intermediate bearing plate which carry the starter drive, oil-pump drive, magneto drive, and generator drive, if replaced, should be located accurately so that all gear clearances are correctly maintained. These bushings should not be replaced unless special equipment is at hand. It is better to install a complete new assembly, sending the old assembly to the factory for reconditioning, rather than have faulty installation.

Accessory drive clearances should have all nicks and burrs smoothed

173

with a fine stone. Nicks, burrs, and scores should be removed from bearing surfaces by using a fine stone or crocus cloth.

If the aluminum plug in the generator or magneto gear is loose, it should be replaced by a cork plug. To replace a cork seal, the inside surface of the gear shaft should be thoroughly cleaned. The inside of the gear shaft is then coated with shellac or a similar alcohol preparation and allowed to dry. The proper-sized cork is then dipped into alcohol, pressed into the place, and allowed to dry. New aluminum plugs should not be inserted if they tend to distort the surface of the shaft. All cork plugs should be replaced with new corks at each overhaul.

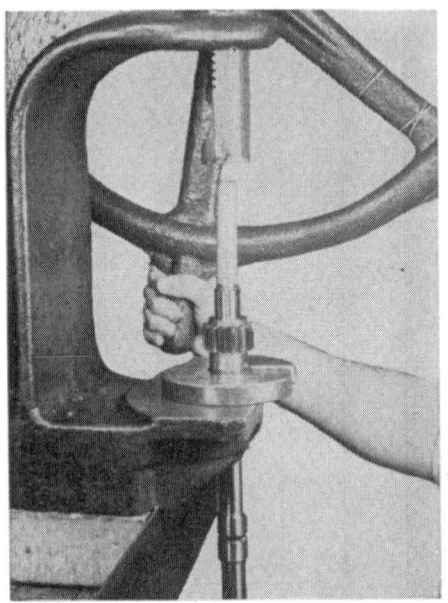

Fig. 131. Replacing a cup plug in the end of a shaft by means of an arbor press. (Courtesy Ranger Aircraft Engines)

Any nicks or burrs should be removed from the rear case parting surface or other parts, and studs replaced where necessary. The various drive-pad surfaces should be smooth and checked for flatness.

If the cylinder barrels have light or moderate scores which may be removed by stoning or the use of wet-or-dry paper, an area 3 or 4 in. wide should be smoothed. Smoothing a narrow strip may allow blow-by. Not more than 0.0015 in. of material should be removed. If it is necessary to remove more than this amount of material, the cylinder should be replaced or bored oversize. It is not usually recommended to remove scoring from a cylinder by honing. If cylinders are honed, they should always be removed after the run-in to check their condition. It is not necessary to remove scratches entirely, so long as the surface close to the crack is stoned so carefully that no sharp edges exist.

Cylinder grinding should only be done with special equipment. It is better practice to send cylinders that require grinding to an approved repair station or the manufacturer. When the grinding is done, choked cylinders should be ground straight.

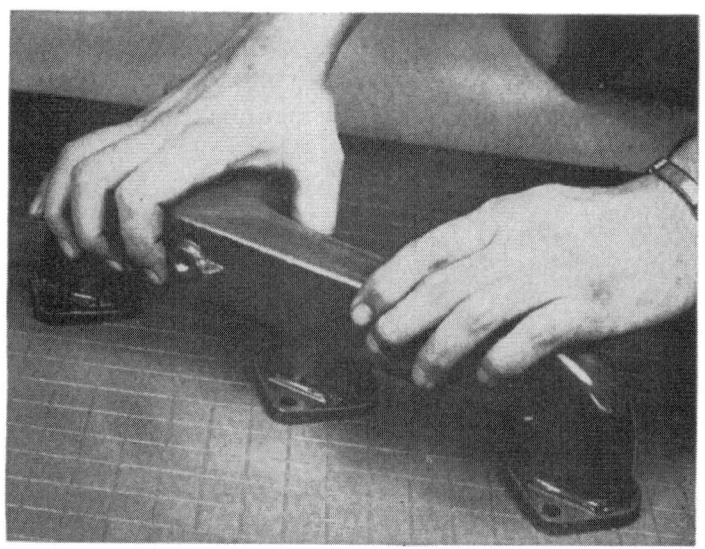

Fig. 132. Intake-manifold-port flanges may be reconditioned by lapping. (Courtesy Ranger Aircraft Engines)

Fig. 133. Installing a rocker-arm bearing by means of an arbor press. (Courtesy Jacobs Aircraft Engine Company)

If the valve guides are to be retained, they should be cleaned of all carbon deposits and any slight rough spots smoothed with crocus cloth and unleaded gasoline. If it is necessary to replace the valve guides, special equipment is necessary and the manufacturer's directions should be carefully followed.

Fig. 134. Pressing in valve guides by means of an arbor press. (Courtesy Jacobs Aircraft Engine Company)

Spark-plug bushings may be removed and replaced in accordance with the manufacturer's instructions, using the tools and methods recommended.

The rocker-box insert may be replaced. The threads in the cylinder head should be cleaned and the new rocker-box insert replaced, using the tools recommended by the manufacturer.

Exhaust-port inserts may be replaced by following the manufacturer's directions and using the proper tools. Valve seats should be refaced at every overhaul. The carbon and glaze should be removed from the seat, and a tool having the proper bevel on the face should be used to produce a new seat. Most valve seats are cut to a 45° angle, although in some engines the valve seat may be at a 30° angle. Some engines require that the valve seat and valve face have a slight angle of approximately $1\frac{1}{2}$° difference between the valve seat and the valve face to allow for a wedging action when the valve closes. The valve seat should never ex-

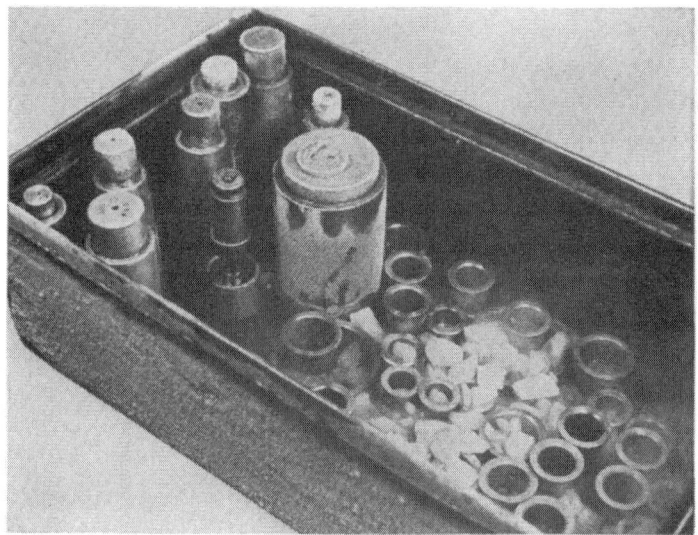

Fig. 135. Chilling engine parts with alcohol and dry ice. (Courtesy Ranger Aircraft Engines)

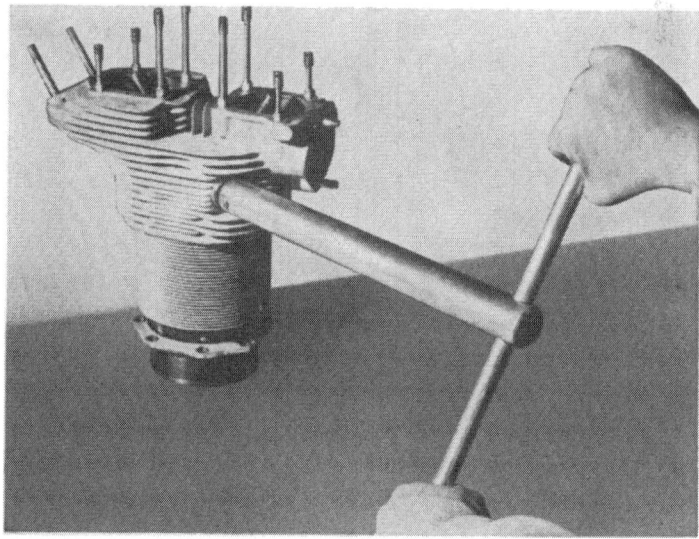

Fig. 136. A new spark-plug insert screwed into the heated cylinder head after having been chilled in a bath of alcohol and dry ice. (Courtesy Ranger Aircraft Engines)

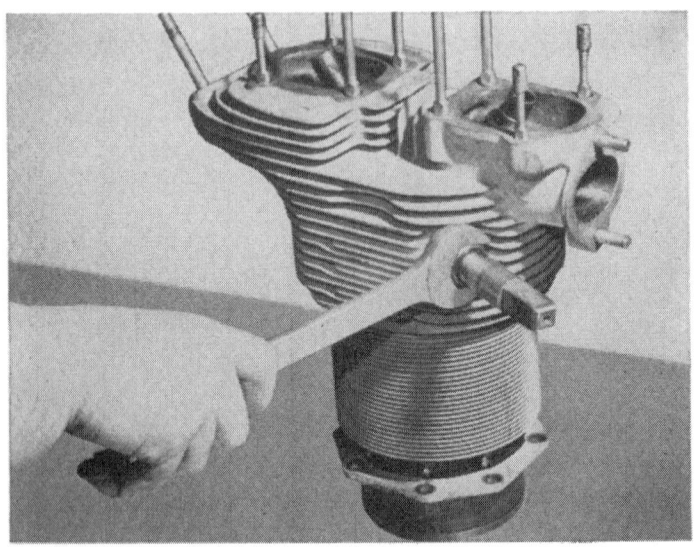

Fig. 137. The driving nut should be loosened before removing the driver. (Courtesy Ranger Aircraft Engines)

Fig. 138. The spark-plug insert is pinned in place by two brass lock pins. The lock-pin holes are drilled 90° apart by the use of a jig and a drill of the proper size. (Courtesy Ranger Aircraft Engines)

ceed the valve face in width. If necessary, the edge of the valve seat should be beveled off so that the seat is less than the width of the face of the valve. All valves should be carefully repaired, having any carbon removed and rough spots smoothed off. Any pitting at the neck of the valve is cause for its rejection. The valves and valve seats should be lapped together forming a continuous fit around the seat and valve face.

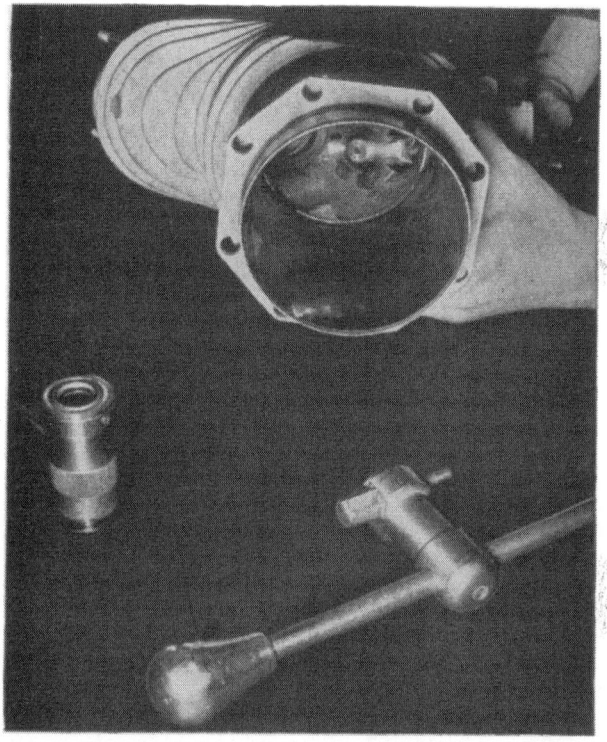

Fig. 139. Refacing a valve seat. (Courtesy Ranger Aircraft Engines)

A light coat of grinding compound should be applied to the valve face. After the valve is inserted in its proper guide, it should be oscillated back and forth with a twisting movement of the wrist, if a hand-grinding tool is used. The valve seat should be checked with Prussian Blue or by inserting the valves and valve springs and testing with the cylinder in the inverted position by pouring gasoline into the valve ports. Care should be taken not to allow any of the grinding compound to collect in the valve guide or on other parts of the cylinder, particularly the cylinder walls. After lapping or grinding the valves the entire cylinder assembly should be thoroughly cleaned to remove all traces of grinding

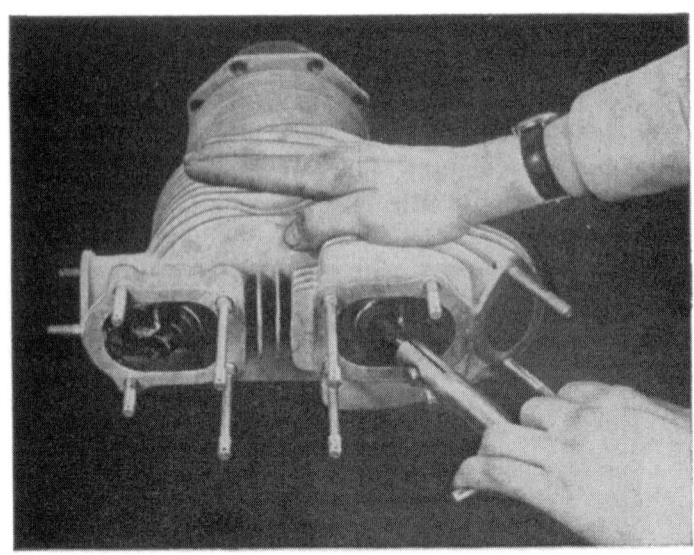

Fig. 140. Lapping valves and valve seats. (Courtesy Ranger Aircraft Engines)

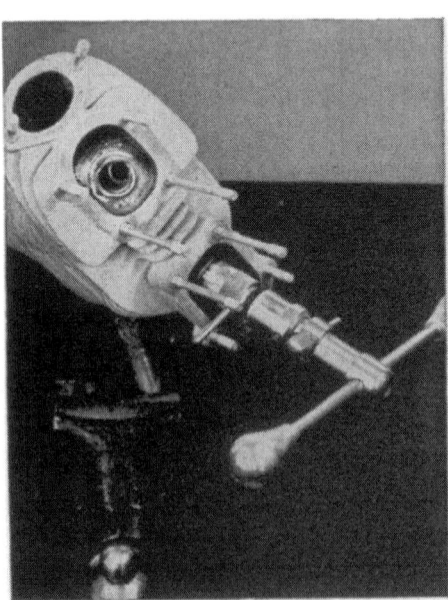

Fig. 141. Refacing an exhaust-valve seat (Courtesy Ranger Aircraft Engines)

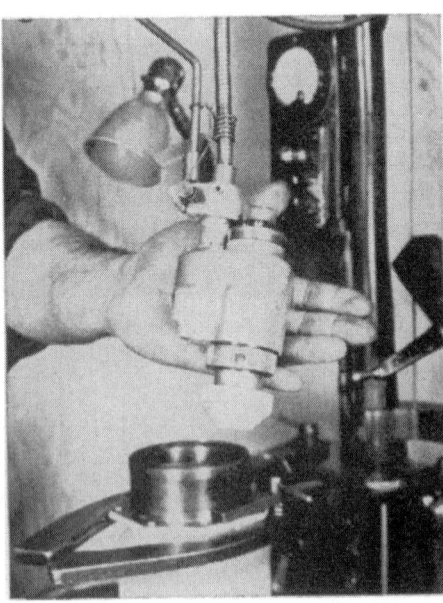

Fig. 142. A grinder used for refacing valve seats. (Courtesy Ranger Aircraft Engines)

180

compound. If the valve seats are damaged to such an extent that they must be replaced, the manufacturer's directions for removing the old seat and inserting the new one should be followed.

Damaged rocker-arm bearings may be replaced by pressing out the old bearing on an arbor press. The new bearing is pressed into place in

Fig. 143. Drilling out the rivet which holds the valve-rocker roller in place. (Courtesy Ranger Aircraft Engines)

a similar manner. The chamfered side of the rocker-arm bearing should enter the bore first.

Valve stems and guides should be cleaned and oiled before assembly. The valve stems should be inserted in the guides, and a block placed inside the cylinder to prevent the valve from falling out. The valve-spring lower washers are inserted over the valve-stem guides. The inner and outer valve springs are inserted in the lower washers. Be sure that the valve springs are properly seated on the washers. The upper valve-spring washers are placed over the valve stem and on the springs. Care should be taken that these washers are properly seated on the springs. By means of a valve-spring compressor, the valve springs are compressed and split locks installed. Care should be taken that the locks are properly seated before releasing the valve-spring compressor. If the split locks are not properly installed in the grooves, the valves will not be properly locked and damage will result.

The rocker arms are inserted in the rocker-box housing and tapped into place. Screw the nuts on the rocker shafts and tighten in accordance with the torque limits given in the table of limits.

It is good practice to replace all piston rings at each overhaul. The piston skirt should have any scores or scratches smoothed by the use of a soft stone. Nicks and burrs should be carefully removed from piston-

Fig. 144. Replacing the rivet through the valve-rocker roller. The rivet head is formed with a punch. (Courtesy Ranger Aircraft Engines)

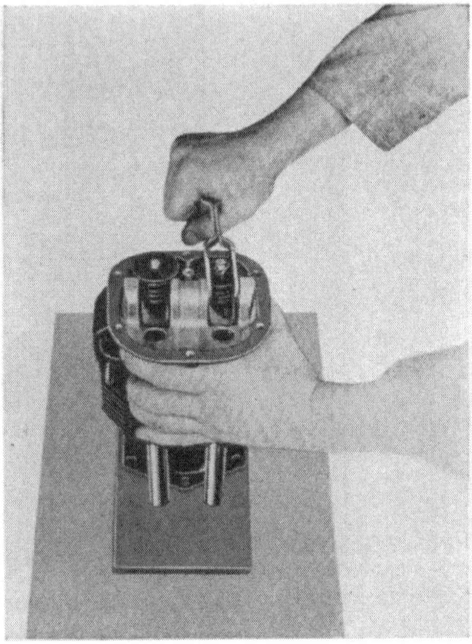

Fig. 145. Compressing a valve spring. (Courtesy Continental Aircraft Engines)

ring grooves. All carbon should be removed from the grooves, taking care not to damage the rounded joint between the bottom of the groove and the land wall.

When a new piston is to be installed, it is important that its weight be within a few tenths of an ounce of the weight of the old one. The weight of each piston is usually found stamped on the top of the piston. Most pistons not only have the weight stamped on the top, but have the part number and the number of the cylinder in which it has been installed.

The diameter of each piston-pin hole to which a new plug must be fitted should be carefully measured. The plug should be turned down to the outside diameter to obtain no more than $\frac{1}{1000}$ in. tight fit in each individual hole. As a rule, the plug should not be allowed to bottom in the hole in the pin. The plug should be placed in dry ice before inserting, or the pin may be heated to a temperature of from 350° to 375° F. in an oil bath. If the pins are heated in an oil bath, care should be taken to remove all traces of oil from within the pin before the plugs are installed.

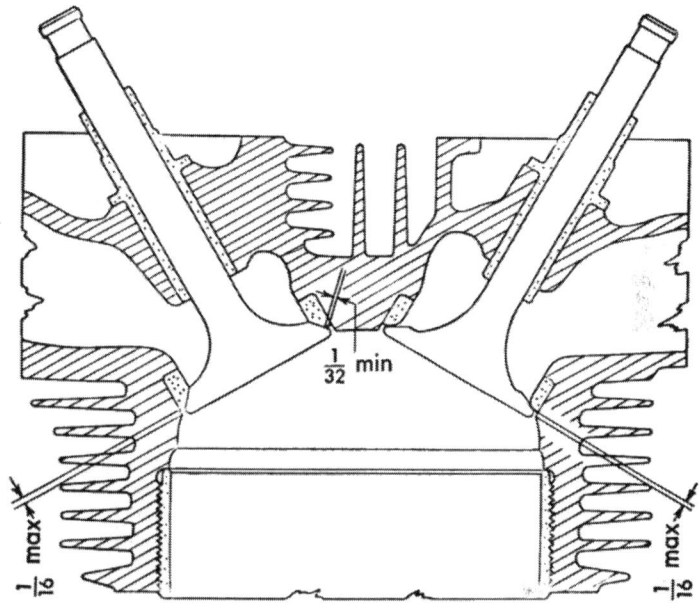

Fig. 146. Valve seats should be cut according to the manufacturer's specifications. (Courtesy Ranger Aircraft Engines)

When installing new rings at overhaul, side and end clearance should be within the table of limits. Oversize rings should be checked for end clearance, using an oversize cylinder barrel. All piston rings must be installed in their proper ring groove in the piston, and the side marked "top" installed toward the piston head or crown. The rings should not be expanded more than necessary when placing them in the grooves. Pistons and rings are furnished in oversizes in order that the proper clearance between the piston and barrel and the end clearance of the rings be maintained when placed in an oversize barrel.

When it is found necessary to replace a main-bearing liner, the manufacturer's directions should be carefully followed.

Tappet guides may be replaced, when necessary, but are not usually removed at the overhaul time unless unusual wear or looseness is noted. The tappet guides should be removed and replaced as recommended by the manufacturer.

Burrs, nicks, and rough places may be removed from the teeth of the cam pinion and cam-drive gear, if necessary, by the use of a smooth stone.

When it is necessary to replace the thrust-bearing liner, the manufacturer's directions should be carefully followed.

Cam-Drive Pinion. If the cam-drive pinion and cam-drive gears have to be removed, the manufacturer's instructions should be carefully observed in their reassembly. To reinstall an outer oil-seal sleeve, it is usually necessary to heat the case from 200° to 250° F.

Crankshaft. Light scratches or gall marks on the crankpin or crankshaft should be removed with crocus cloth or wet-or-dry paper. It is allowable to use a stone lightly, if necessary, before using the crocus cloth. The clamping area on the crankpin should not be stoned any more than absolutely necessary. Any reduction of the pin diameter or uneven spots on its surface will produce a poorer clamping action.

All oil passages within the crankshaft must be thoroughly cleaned.

Fig. 147. Replacing the starter jaw in the rear end of the crankshaft. The crankshaft has been heated to approximately 450° F. (Courtesy Ranger Aircraft Engines)

Most crankshafts are equipped with threaded plugs which may be removed to clean the oil passages. It is not usually necessary to remove any of the plugs which are permanently pressed in. If any of the pressed-in plugs have become loosened or shifted from their original position, they must be replaced. These plugs are available in oversizes.

When replacing a plug of this type, it must be hand-fitted to the clearance given in the table of limits. After plugs have been fitted and pressed in, the outside of the crankshaft at the plug location should be carefully checked to make sure that the plug has not expanded the shaft. If a slight bulge has been caused by the pressing in of the plug, the

Fig. 148. To prevent the crankshaft's being forced out of alignment when replacing the starter jaw, it should be properly braced. (Courtesy Ranger Aircraft Engines)

crankshaft must be stoned down to its true diameter. If threaded plugs are staked into place, it is usually necessary to remove small burrs left from drilling the stake hole and removing the plug.

Most manufacturers recommend that the old method of locking plugs by staking be converted to locking the plugs with a lock, lock screw, and safety wire. The manufacturer's directions for this operation should be followed. The manufacturer's manual should be carefully followed in the assembly of the crankshaft.

Master and Link Rod Assembly. All oil passages in the master rod must be open and cleaned. Burrs or gall marks in the knuckle-pin holes of the master rod should be carefully removed with a soft stone. All scores and burrs must be removed from the knuckle pin by light stoning. Any stoning should be kept to a minimum so as not to destroy the seat of the pin in the master rod. The surface of the master-rod bearing should not be disturbed unless it is absolutely necessary. A fine stone or crocus cloth may be used to remove very slight irregularities. The bearings

185

should under no circumstances be scraped or burnished. Master rods or link rods should never be bent to attain proper alignment.

Most link rods are furnished with pressed-in bushings, but some link rods are of heat-treated aluminum alloy and are unbushed. These aluminum link rods are usually replaced at every second overhaul.

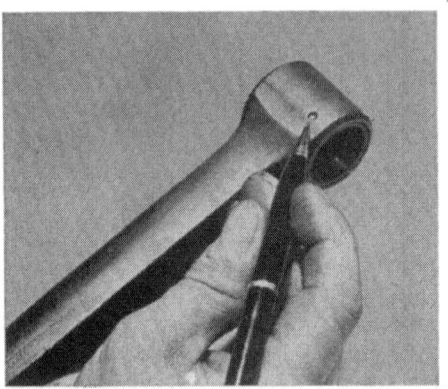

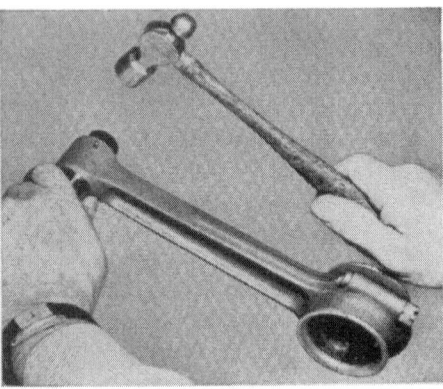

Fig. 149. The pin retaining the bushing in the connecting rod must be removed before removing the bushing. (Courtesy Ranger Aircraft Engines)

Fig. 150. Replacing the bushing retaining pin. (Courtesy Ranger Aircraft Engines)

Fig. 151. Installing knuckle pins by means of an arbor press. (Courtesy Jacobs Aircraft Engine Company)

REPAIR AND REPLACEMENT

When rebushing link rods, it is customary to replace the bushings in both ends to obtain proper alignment. The rebushing should be done in accordance with the manufacturer's manual.

Master rods with badly damaged bearings should usually be returned to the manufacturer for replacement. If the proper facilities are avail-

Fig. 152. Replacing master rod bearing by means of an arbor press. (Courtesy Jacobs Aircraft Engine Company)

able for replacement and reboring of the master-rod bearing, the manufacturer's instructions should be carefully followed. The assembly of the master rod and link rods should be in accordance with the manufacturer's directions. It is absolutely essential that all safety devices be installed properly in all internal parts of the motor at the time of assembly.

XIII ENGINE ASSEMBLY

The proper operation of the engine depends to a great extent upon its being properly assembled. Most assembly operations must be performed with an exactness which can only come through experience and training. More than any other place in the mechanical field, the aircraft

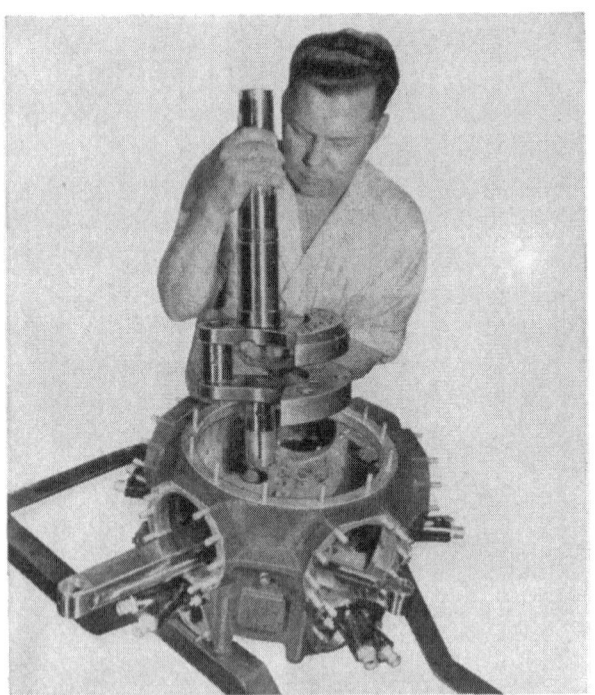

Fig. 153. Installing a crankshaft. (Courtesy Kinner Motors, Incorporated)

engine mechanic must be familiar with close tolerances, exact fits, and compliances with standards.

The standards for the assembly of an aircraft engine are exacting throughout. The mechanic should be thoroughly familiar with the table of limits and should follow the manufacturer's assembly directions with the greatest possible care. Dirt, dust, and grit are perhaps the greatest enemies that the engine has. Assembly, insofar as possible, must be made in a dust-free room.

The proper tools should always be used during assembly, and ex-

Fig. 154. Installing one half of a split crankcase. (Courtesy Continental Aircraft Engines)

treme care must be taken to avoid damaging the bearing surface of any part which is finished to close tolerances. Microscopic cracks or scratches on such surfaces may lead to engine failure.

All gaskets, packings, hose, and leather or rubber seals should be replaced at the overhaul period. Gears, shafts, and bearings should receive an adequate coating of oil just prior to installation. All safety devices, such as lock nuts, flanged washers, lock washers, cotter pins, and safety wire, should be installed as soon as the part to which they are attached is in place, unless a safety device includes more than one item, such as a safety wire passing through several cap-screw heads.

Each bolt, nut, or other part should be safetied at once. Do not leave a series of installations to be safetied at a later time. They may be overlooked.

Crankcase. The crankcase, which is the main supporting element of the engine, is usually the first part to be considered when assembling an engine. Some part of the crankcase is usually fastened to the assembly stand. The fastening should be securely made by means of the proper mounting lugs. In a radial engine, the rear main section of the crankcase is usually mounted to the assembly stand. The bearing races and assembled rollers should already be in place in the crankcase section.

Crankshaft and Accessories. The rear crankshaft on a split shaft, or the whole crankshaft if it is in one piece, is usually next installed. On some engines, the crankshaft and the connecting rods are installed as a

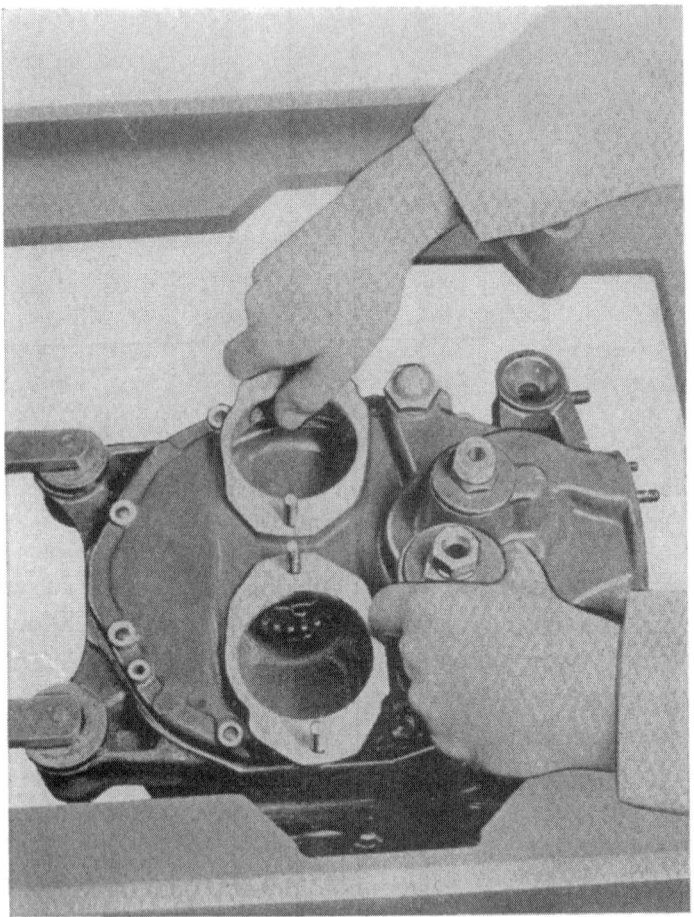

Fig. 155. Installing a crankcase cover. (Courtesy Continental Aircraft Engines)

subassembly. The crankshaft bearings should be guarded carefully against any damage. When parts need tapping into place, a soft mallet should be used. This mallet should be of wood, plastic, rubber, raw-hide, lead, or some other soft material which cannot possibly damage the part.

The cluster gears and accessory drive gears which are attached to the

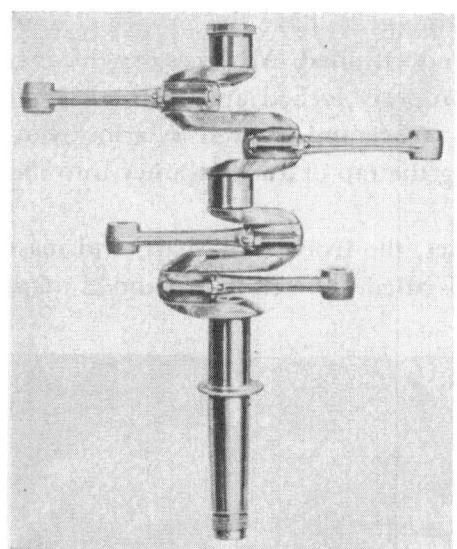

Fig. 156. Assembly of connecting rods on a crankshaft. (Courtesy Continental Aircraft Engines)

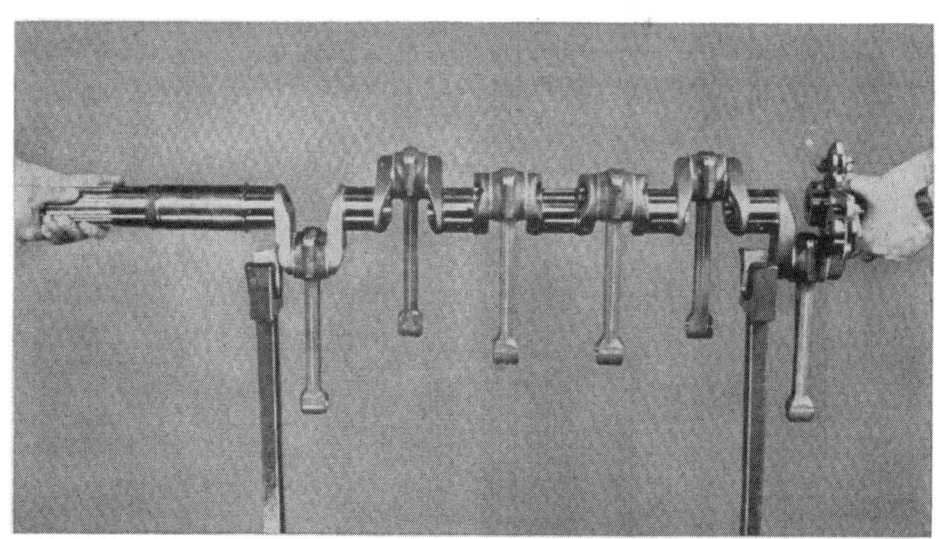

Fig. 157. The crankshaft with connecting rods attached. (Courtesy Ranger Aircraft Engines)

rear crankshaft are usually installed next. A check should be made to see that the matching gears have the proper clearances and that their mating surfaces are well oiled. All accessory drive gears, shafts, and so forth should be properly locked and safetied. Items of this kind are often fastened into place and secured by a lock nut which is, in turn, secured by bending the tab of a lock washer into the proper slot of the lock nut.

On radial engines, the front crankshaft, and master-and-link-rod assembly are next installed. The crankpin and crankpin hole in the rear

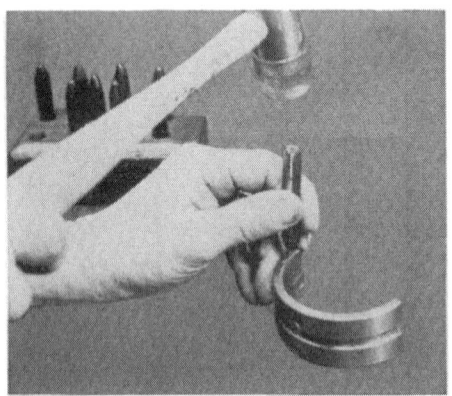

Fig. 158. New bearings should always be stamped with the proper location number. (Courtesy Ranger Aircraft Engines)

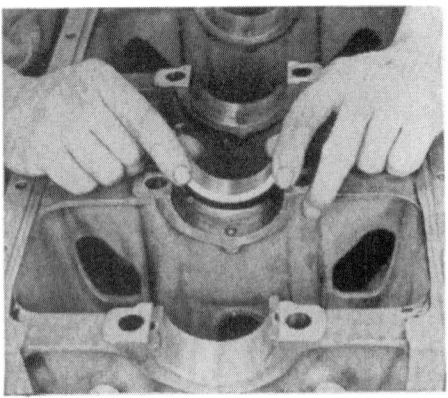

Fig. 159. Placing the lower half of a main bearing in the crankcase. (Courtesy Ranger Aircraft Engines)

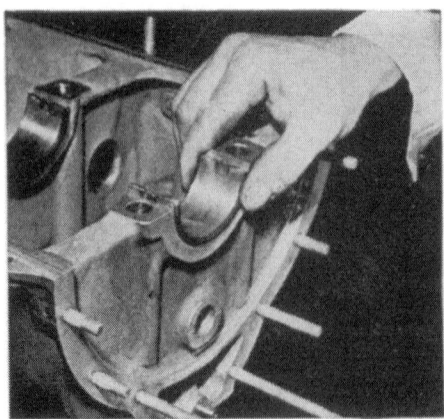

Fig. 160. The lower half of the main bearing in place in the crankcase. (Courtesy Ranger Aircraft Engines)

Fig. 161. Installing a connecting rod. (Courtesy Ranger Aircraft Engines)

crankshaft cheek should be cleaned with carbon tetrachloride to remove any oil which may remain on these parts. It is essential that these surfaces be kept absolutely clean until assembled and clamped tight.

The master-rod bearing should be wiped clean of oil. The master rod with the link rods in place is carefully placed in position, and the

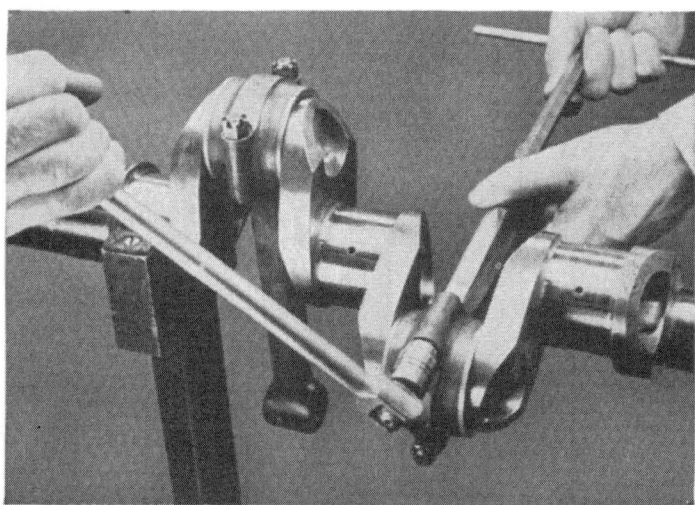

Fig. 162. Tightening the connecting-rod bolts. (Courtesy Ranger Aircraft Engines)

Fig. 163. The side clearance of the connecting rods should be checked with the proper feeler gauge. (Courtesy Ranger Aircraft Engines)

front half of the master rod is very carefully inserted through the master-rod bearing and into the crankpin hole in the rear half of the crankshaft. Usually the crankshaft clamp must be spread slightly with a proper wedge to allow easy entry of the crankpin.

The aligning bar should be placed through the aligning holes in both

Fig. 164. Installing the front half of a crankshaft. (Courtesy Jacobs Aircraft Engine Company)

the counterweights. The wedge should be removed from the clamp. A feeler gauge should be inserted between the master rod and the front cheek of the crankshaft to ensure the required end clearance.

The length of the crankpin clamp bolt should be measured with a micrometer before being put into place. The crankpin clamp bolt should be inserted from the proper side as directed by the manufacturer. Inserting the pin from the wrong side may destroy perfect balance. After the clamp bolt is in place, the threads and face of the clamp-bolt nut should be lubricated lightly with the proper oil. Before tightening the clamp bolt, the alignment of the crankshaft must be rechecked, making sure that the aligning bar will enter the holes in both counterweights without binding. The aligning bars should be withdrawn from the lower counterweight before tightening the bolt. If left in place, the aligning bar might be bent slightly and bound in the hole. The clamp bolt should then be tightened with the proper wrench until it is stretched the amount given in the table of limits. Care should be taken

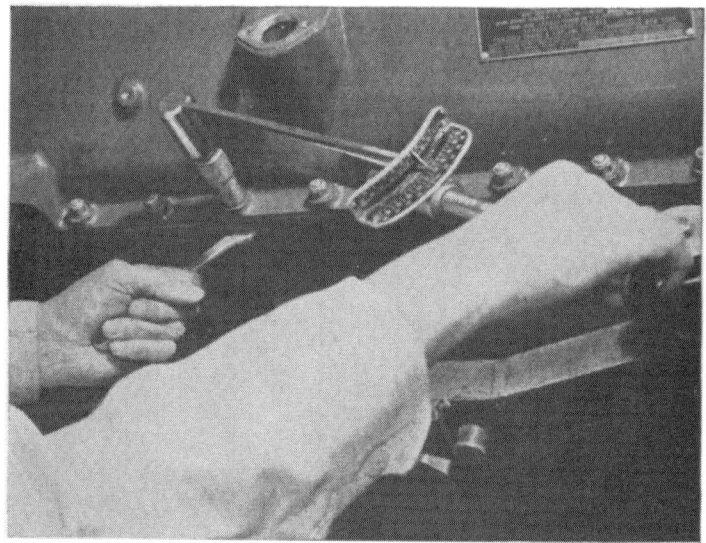

Fig. 165. The crankcase flange nuts should be tightened with the proper torsion wrench. (Courtesy Ranger Aircraft Engines)

to measure this stretch accurately. After the bolt has been measured and the proper stretch has been obtained, it is good practice to prick-punch the bolt and nut so that, in case the nut has to be loosened, it can be brought back to the proper point by lining up the punch marks.

The aligning bar should again be used to check the alignment of the crankshaft. If the bar does not enter the aligning holes easily, the crankshaft is not in proper alignment. In this case, the clamp-bolt nut must be loosened and the aligning procedure repeated. The feeler gauge determining the bearing end clearance must remain in place until the clamp bolt is firmly tightened. When the crankshaft is properly aligned and the clamp bolt stretched the proper amount, the cotter pin is placed through the clamp bolt. The ends should be bent firmly back. The key should not be free to vibrate in the hole. The protecting mask should be removed from the oil holes in the crankshaft, and the shaft wiped clean.

The front half of the main crankcase is usually installed next. It should be placed on the shaft without being tightened into place. The bearings should then be installed on the front crankshaft. After the main roller bearings are in place, and properly aligned, the front crankcase may be tightened into place. In tightening the parts of a crankcase together, all the bolts should be drawn into place snugly, so

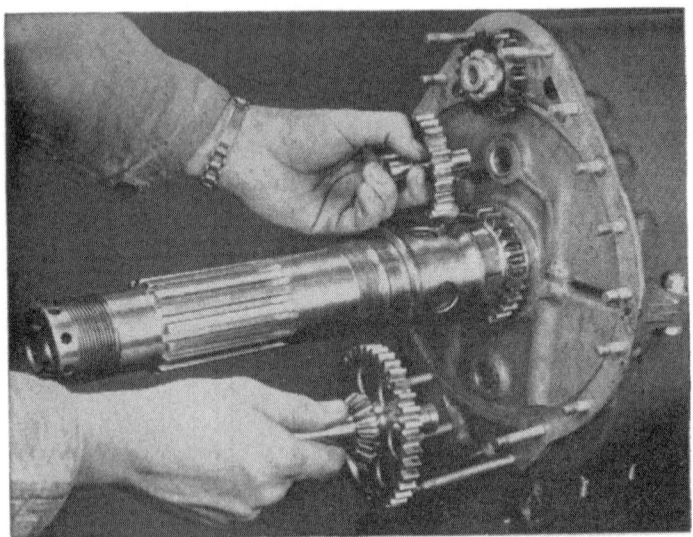

Fig. 166. Installing the upper and lower front idler gears. (Courtesy Ranger Aircraft Engines)

Fig. 167. Installing the accessory drive-shaft rear gear. (Courtesy Ranger Aircraft Engines)

that the case is drawn evenly together. The bolts are then tightened with a proper torque wrench as specified in the table of limits.

At this point in the assembly the crankshaft run-out should be checked by use of a dial indicator. If the aligning bar entered the aligning holes easily and without binding, the shafts are in alignment, but

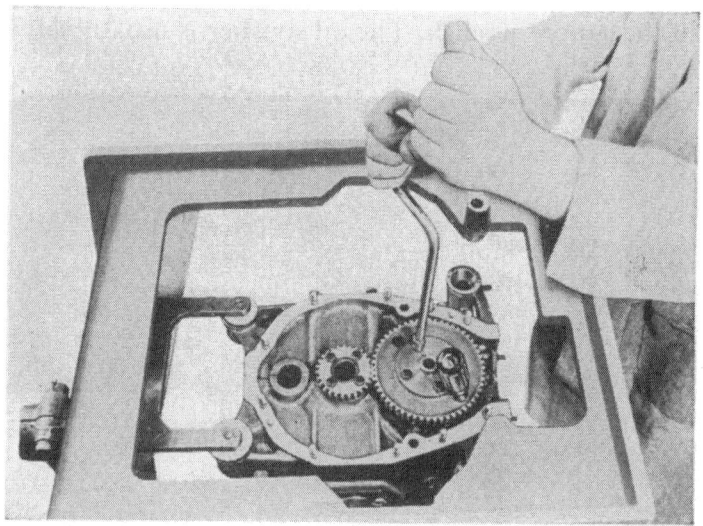

Fig. 168. Installing gears in a crankcase. (Courtesy Continental Aircraft Engines)

if the indicator dial gives a higher reading than that allowed, the crankshaft must be removed and checked for straightness.

Before assembling the cam bearings, cam assembly, or other assembly on the crankshaft, the crankshaft should be coated with white lead, oil, or some similar substance to prevent scratches which might be caused by bringing together two dry surfaces.

Before placing the cam bearings on the crankshaft, a check should be made to see that the retainer plate of the front main roller bearing is properly seated. The plate may be greased lightly to hold it in place. If the plate is not properly located, it may break when the timing gears are driven on. The timing-gear key should be inserted in the crankshaft and the timing gear put in place, tapping lightly to ensure proper seating on the key.

At this point, a radial engine may usually be turned so that the crankshaft is in a horizontal position.

Engine Accessories. The engine accessories, such as the oil-pump assembly, carburetor, and oil-sump strainer, should next be installed on their proper mounting pads, and the crankshaft turned until the pump shaft can properly enter the oil-pump drive gear. The pump is then fastened into place.

The carburetor is placed on its mounting studs, fastened into place, and safetied.

The oil-sump strainer is inserted in the crankcase and tightened into place with the proper wrench. The oil strainer is usually safetied to the

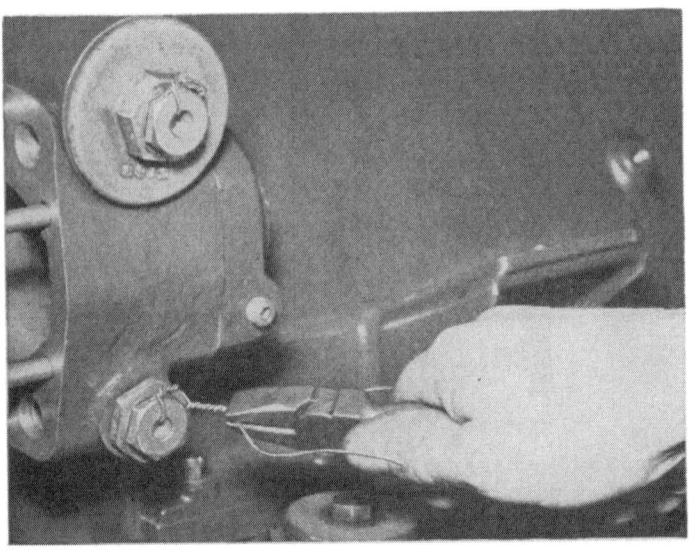

Fig. 169. The oil inlet screen magnet and inlet screen are safetied with wire. (Courtesy Ranger Aircraft Engines)

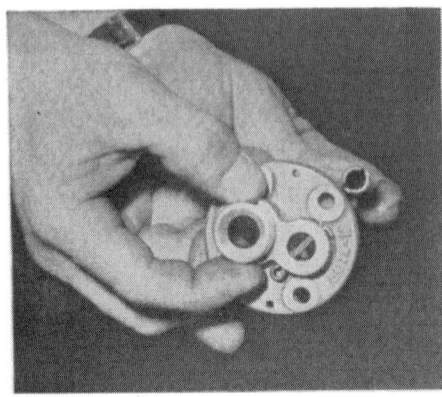

Fig. 170. The pressure oil pump idler gear shaft is secured in the pump cover with a cotter pin. (Courtesy Ranger Aircraft Engines)

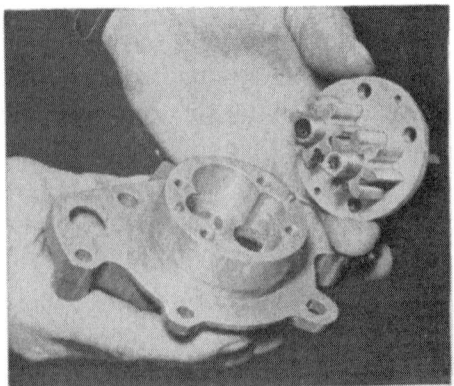

Fig. 171. The assembled parts of the oil-pressure pump are placed in the housing. (Courtesy Ranger Aircraft Engines)

oil-sump plug. Whenever possible at this point, it is well to force oil through the pressure line leading to the crankshaft to lubricate the crankshaft bearing thoroughly and to prevent scoring the bearing surfaces when the crankshaft is turned during assembly. A pressure gun

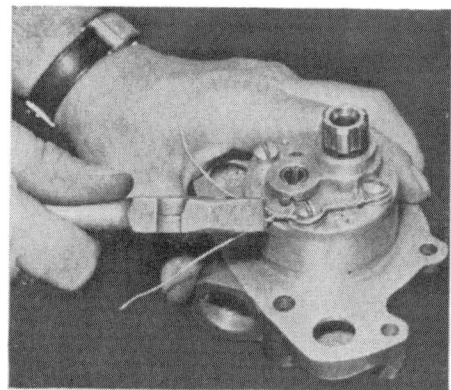

Fig. 172. The cap screws in the pump housing are safetied with safety wire. (Courtesy Ranger Aircraft Engines)

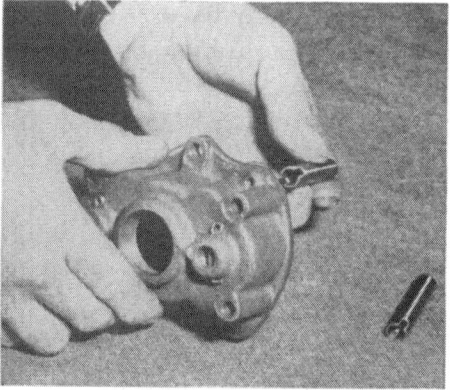

Fig. 173. Scavenge-oil-pump idler gear shaft is secured in the oil pump cover with a cotter pin. (Courtesy Ranger Aircraft Engines)

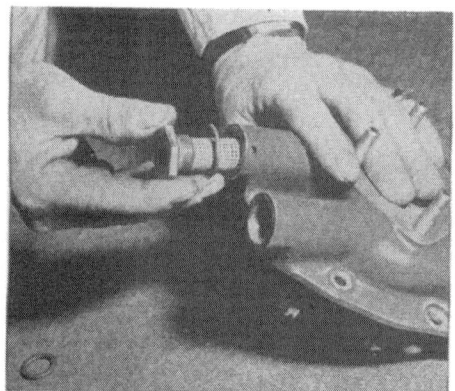

Fig. 174. Installing the finger strainers. (Courtesy Ranger Aircraft Engines)

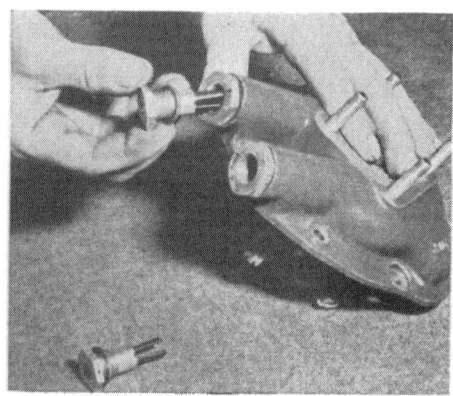

Fig. 175. Installing the magnets in the finger strainers. (Courtesy Ranger Aircraft Engines)

filled with the proper oil may be attached to the pipe fitting. Enough oil should be inserted so that the oil flows freely from the ends of the master-rod bearing. The crankshaft should be rotated to distribute the oil evenly on the bearing. Cuffs are usually placed around the link rods to prevent injury to the link rods or other parts of the engine when the crankshaft is turned. The cuff also protects the link rod from being accidentally bumped against a cylinder hole in the crankcase.

Before installing the piston pins, pistons, and cylinders, all bearing surfaces should be thoroughly lubricated with engine oil. The pistons are then placed over the link rods and the piston pins inserted. Care should be taken to see that the pistons are placed in their proper cylinder. The number of the cylinder into which the piston is to be

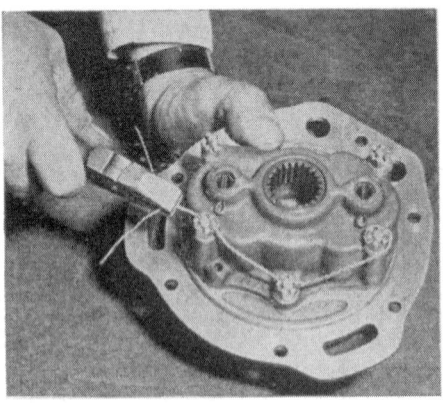

Fig. 176. Safetying the scavenge oil pump attaching nuts with safety wire. (Courtesy Ranger Aircraft Engines)

Fig. 177. Piston rings are installed by using a piston-ring expander. Do not expand the ring more than is necessary to install. (Courtesy Ranger Aircraft Engines)

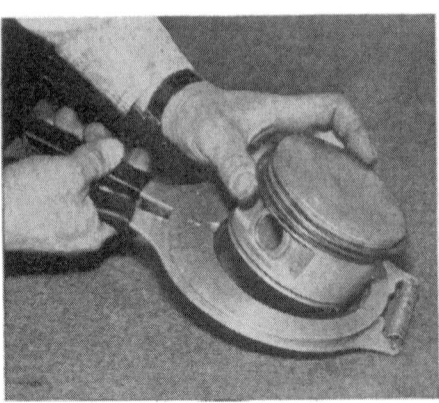

Fig. 178. Most rings are marked "Top" on one side; this side should be installed toward the piston head. (Courtesy Ranger Aircraft Engines)

Fig. 179. Piston rings should be checked all the way around for the proper side clearance by using a feeler gauge of the correct thickness. (Courtesy Ranger Aircraft Engines)

put is usually stamped on the piston crown above the front pin boss. This number is placed toward the front of the engine.

The cylinder wall of No. 1 cylinder which is the master-rod cylinder should be thoroughly lubricated with engine oil before placing over the piston. New cylinder oil-seal rings should be in place on the cylinder skirt. Each ring groove in the piston should be liberally treated with engine oil. The crankshaft, which should now be in a vertical position, should be turned to bring the master rod to top dead center. The cylinder should be slid

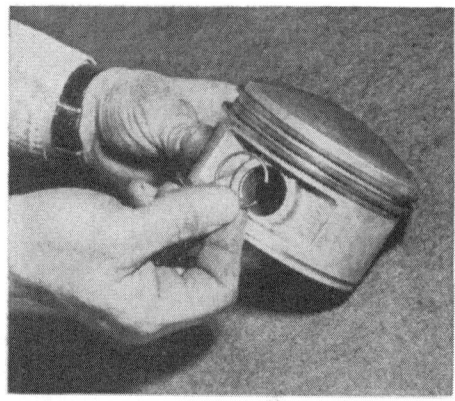

Fig. 180. New circlips should always be installed. (Courtesy Ranger Aircraft Engines)

carefully and squarely into place, using a ring compressor. This procedure should be followed until all the cylinders are in place, attaching each cylinder to the crankcase in turn.

The cylinder hold-down nuts should be tightened snugly all around to make sure that the cylinder mounting flange is evenly pressed against the mounting pads. The nuts should then be tightened to the proper torque with a special hold-down nut wrench. The proper torque is given in the table of limits.

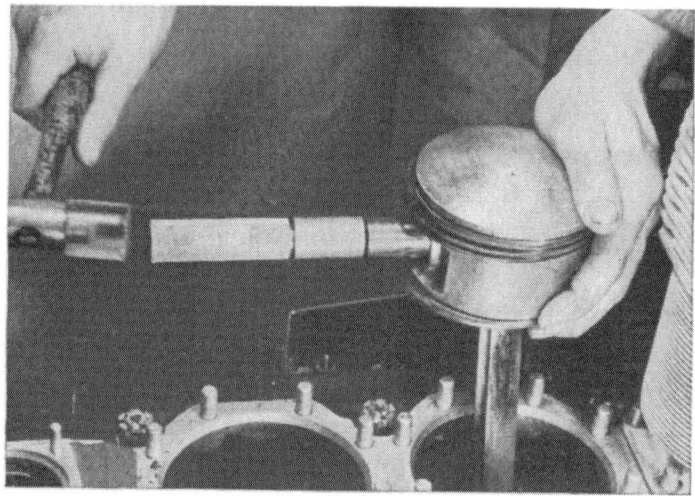

Fig. 181. Installing a piston pin using a plastic mallet and a piston-pin drift. (Courtesy Ranger Aircraft Engines)

201

Intake pipes are usually installed next. The packing nuts, into which new packing has been inserted, are tightened to hold the pipes in place. These nuts should not be too tight as this might prevent the pipe's sliding in and out of the rubber packing gland during the expansion and contraction of the cylinder. If the packing nut is too tight, the result may be cracks in the pipe flange at the cylinder head. With the engine rotated until the crankshaft is in the horizontal position, the front case or front crankcase cover is usually installed next.

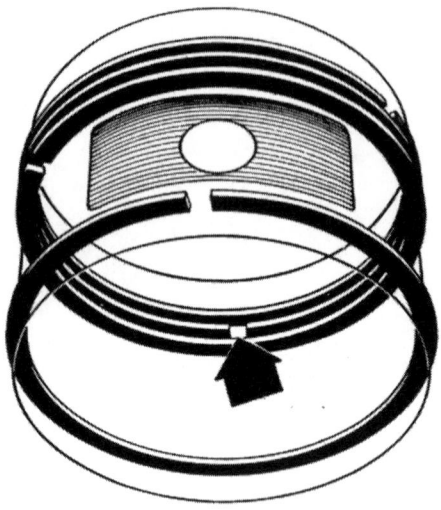

A timing disk should be placed on the crankshaft to indicate the exact top dead center of No. 1 which is the master-rod cylinder. On most spline crankshafts, a blank spline is usually aligned with the crank throw. The zero reading of the timing disk should be in line with this spline. Some engines have an arrow on the end of the crankshaft to indicate the crank throw.

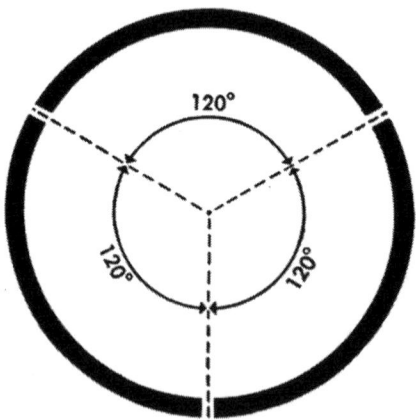

Fig. 182. The ring gaps should be installed at even intervals around the piston. (Courtesy Ranger Aircraft Engines)

A timing wire is usually prepared to keep the cam stationary, the wire being inserted through the timing hole in the cam hub and the front main crankcase. A steel straightedge is placed over the top two cam lobes, and the cam is turned until there is an equal distance between the steel straightedge and the studs directly above. The cam, by means of the wire, should be locked in this position. The cam must not be free to rotate. It may be rocked gently to make certain that it will not get out of position. If the cam is accidentally moved, it must be realigned with the crankshaft.

Before putting the front case into place, make sure that the pins are in the tappet rollers. All top tappets should be held up. This usually may be done by a piece of wire inserted in the oil holes of the adjoining tappets. With the front case in position, adjust the cam pinion gear with the fingers through the timing hole until the tooth with the punch mark or other timing mark is between the two teeth that are marked on the timing gear. If a new gear is being installed, the manufacturer's

Fig. 183. Oiling the piston for cylinder installation. (Courtesy Continental Aircraft Engines)

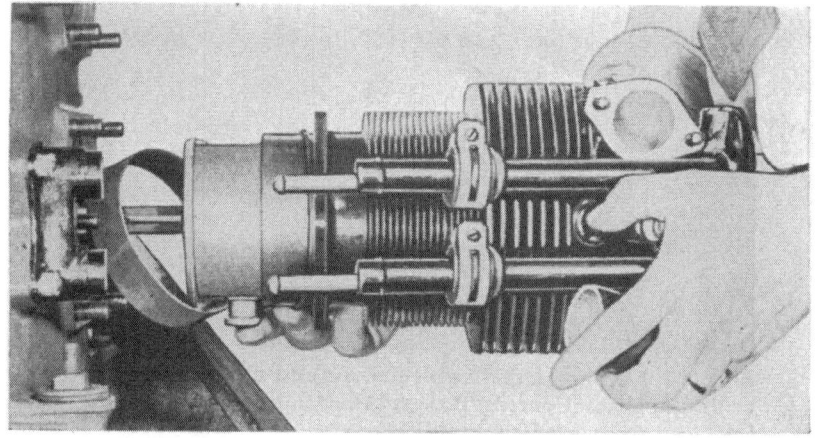

Fig. 184. Installing a cylinder on an opposed engine. (Courtesy Continental Aircraft Engines)

Fig. 185. Installing a cylinder on a radial engine. (Courtesy Jacobs Aircraft Engine Company)

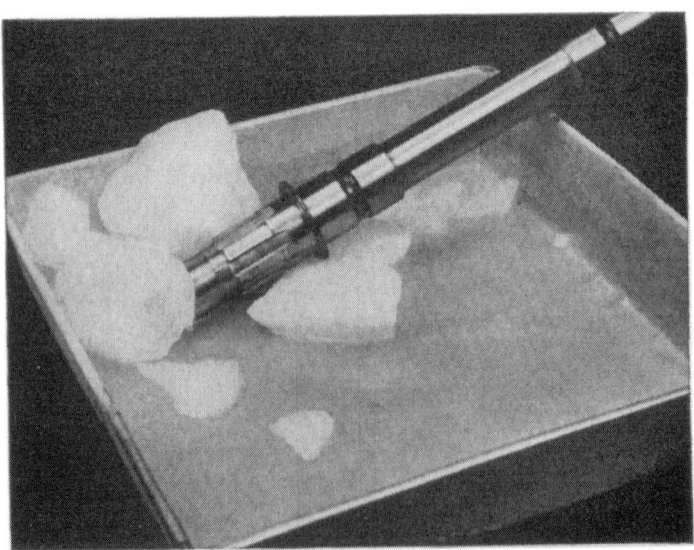

Fig. 186. Chilling a shaft with dry ice and alcohol before installing a gear. (Courtesy Ranger Aircraft Engines)

204

directions for locating the proper tooth to be used in timing should be followed. The front case is then tightened into place.

If the case does not slide into place easily, make sure that the tappets or tappet rollers are not interfering with the cam lobes. Before giving the nuts on the studs their final tightening, check the timing to see that

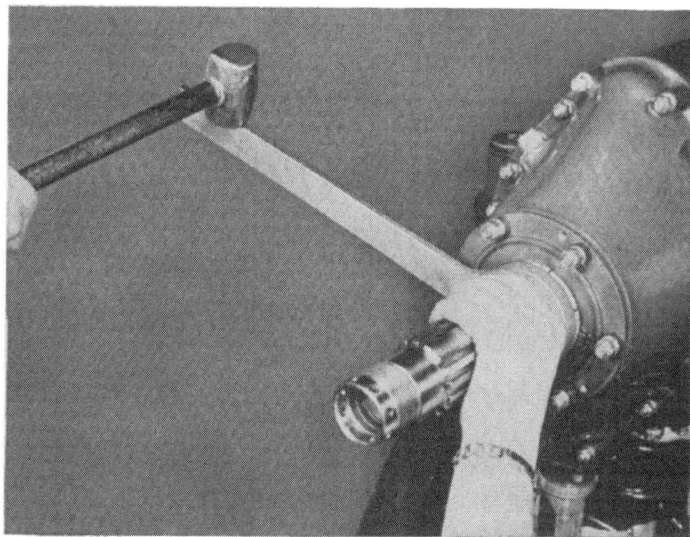

Fig. 187. Tightening the thrust-bearing nut by tapping the wrench with a lead mallet. (Courtesy Ranger Aircraft Engines)

Fig. 188. Installing the thrust bearing on the crankshaft. (Courtesy Ranger Aircraft Engines)

it is correct. The valve timing should be carefully checked, following the manufacturer's directions for each particular engine.

The magneto and the distributor are installed in accordance with the manufacturer's instructions. The oil seals, lines, thrust bearings, thrust plate, thrust nuts, and other parts are then installed. The manufacturer's instructions for the engine should be followed.

Fig. 189. Installing the oil slinger ring. (Courtesy Ranger Aircraft Engines)

Fig. 190. Installing the cover and shim on the front crankcase. (Courtesy Ranger Aircraft Engines)

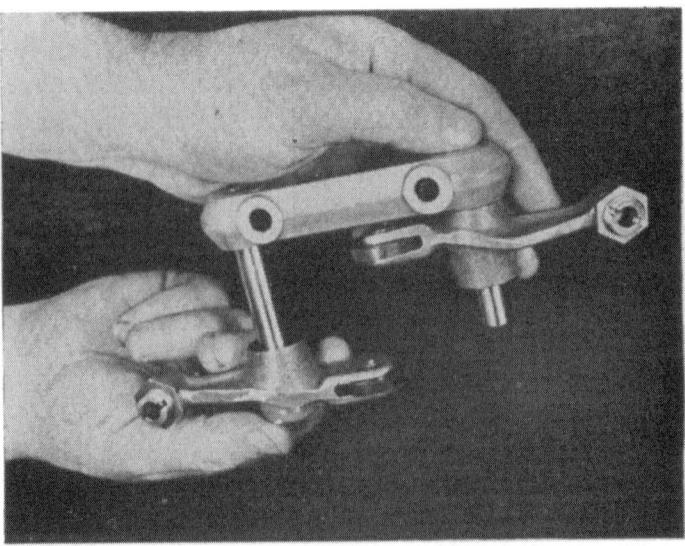

Fig. 191. Installing a rocker arm on its shaft. (Courtesy Ranger Aircraft Engines)

The oil-sump drain hose and priming system are installed next. The push-rod cover tubes, push rods, rocker covers, rocker seal tube, rocker *scavenger* lines, ignition manifold, and any other parts should then be installed. These are all installed by following the directions given in the manufacturer's manual.

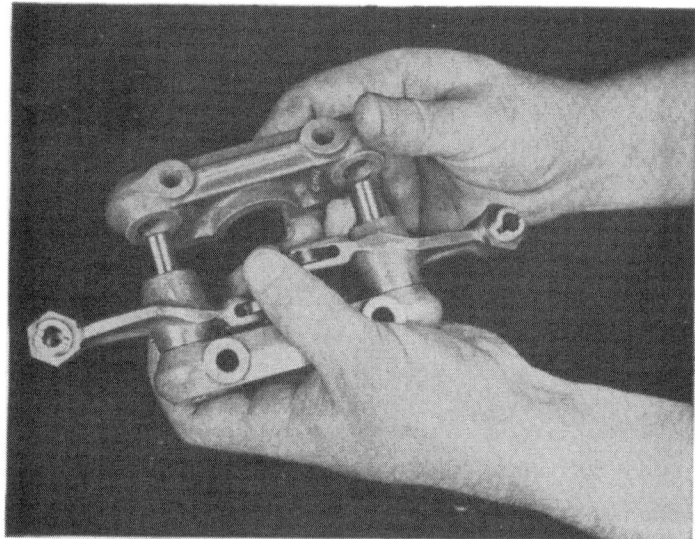

Fig. 192. Installing the second valve-rocker support. (Courtesy Ranger Aircraft Engines)

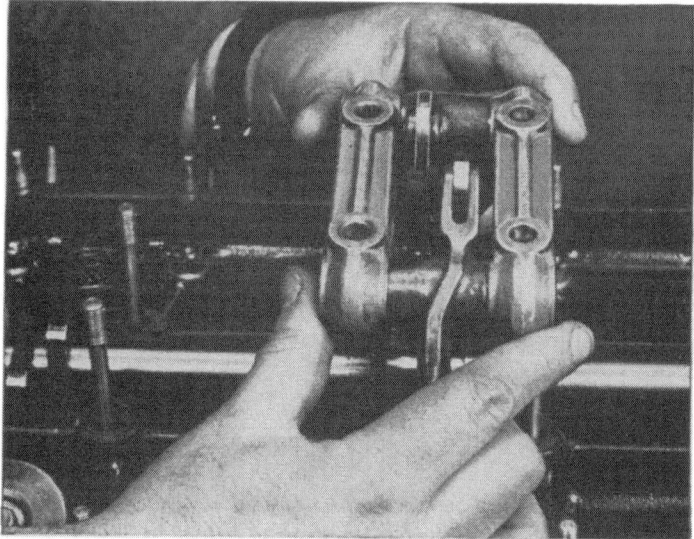

Fig. 193. Rocker-arm assembly unit. (Courtesy Ranger Aircraft Engines)

The procedure suggested in this chapter should be adapted to the engine on which the mechanic is working. An engine should never be assembled without having at hand the manufacturer's manual for that particular engine.

Fig. 194. Checking the rocker-arm clearance. (Courtesy Ranger Aircraft Engines)

Fig. 195. Tightening the push-rod cover-tube packing nut. (Courtesy Jacobs Aircraft Engine Company)

IV ENGINE INSTALLATION, RUN–IN, AND STORAGE

The engine comes from the factory packed in a special case. The bolts
[h]lding down the cover should be removed and the cover taken off with
[car]e not to damage the engine. The tool kit and engine accessories are
[str]apped to the bottom of the packing case.

The thread-protecting cap should be removed from the propeller
[sh]aft. A propeller-shaft lifting eye should be attached to the propeller
[sha]ft, in case of a radial engine, and the engine hoisted by means of a
[cha]in hoist attached to the eye, or a lifting sling should be attached ac-
[cor]ding to the manufacturer's directions.

The engine should be mounted on an assembly stand in such a posi-
[tio]n that the cylinders will drain of rust-preventive material when the
[du]mmy spark plugs are removed. The engine should be rotated several
[tim]es to expel any excess rust preventive in the combustion chambers.
[Th]e dummy plug should then be replaced to prevent foreign matter
[fro]m entering. The engine should then be placed in such a position as
[to] allow the oil sump to be drained of any rust-preventive material.

The Cuno oil filter should be removed and dipped in gasoline with-
[ou]t rotating the strainer. The strainer should then be blown out with
[com]pressed air. After the filter has been thoroughly cleaned, it should
[be] dipped into clean engine oil before installation in the engine. All
[oi]ls should be removed from the exhaust ports and other openings.

In moving the engine to place it on the engine mount or engine
[sta]nd, care should be taken that no parts of the engine come into con-
[tac]t with parts of the stand or engine mount which might cause damage.
[Be]fore removing the engine from the stand, all accessories which will
[no]t interfere with the removal from the stand should be attached. Care
[sho]uld be taken not to damage the two high-tension terminals on the
[ou]tside of each magneto when mounting the engine. The engine should

be raised to a level which will permit the passage of the rear section through the engine mount or which will allow the correct alignment with the engine mount.

Fig. 196. Proper method of lifting a radial type of aircraft engine. (Courtesy Jacobs Aircraft Engine Company)

Engine Installation. To install the engine, the following should be performed in the order given.

1. Fasten the mounting studs using the proper bolts, nuts and washers.

2. Connect the throttle rods at the rear of the carburetor.

3. Connect the mixture-control rods at the right side of the carburetor.

4. Connect the hot-air control rod.

5. If one is used, the starter-adapter drain line should be connected between the bottom of the starter adapter and the oil sump.

6. Connect the oil inlet pipe to the pump.

7. Connect the oil outlet pipe to the pump.

8. Connect the oil-pressure-gauge pipe.

9. Connect the oil-tank vent pipe.

10. Connect the fuel line to the carburetor.

11. Connect the fuel supply pipe to the fuel pump.
12. Connect the fuel-pressure-gauge line.
13. Connect the fuel priming supply pipes.

Fig. 197. Hoisting an engine with a lifting sling. (Courtesy Continental Aircraft Engines)

14. Connect the magneto-wiring ground connections. Be sure the high-tension terminals are tight and in place.
15. Connect the engine-ground bonding wire to the airplane frame.
16. Connect the thermocouple wire, when used.
17. Connect the tachometer cable.
18. Connect the exhaust manifold and pipes.
19. Connect the carburetor air connections.
20. Connect all other accessories and pipes or wiring.
21. All sections of cowling should be installed where necessary.

Before installing a propeller or propeller hub, all parts should be examined for damage or defects and checked for proper fitting. Any corrosion or rough spots on joining surfaces should be carefully dressed off. All parts should be thoroughly cleaned before the propeller or hub is assembled on the shaft. All internal surfaces, except the threaded portion of the hub-retaining nut, should be coated with clean engine oil to provide lubrication and to prevent corrosion. The threaded portion of the hub-retaining nut should be coated with a special thread lubricant. Make certain that the rear face of the rear cone will bear against the thrust-bearing lock nut without interference between the inside chamfer on the cone and the fillet on the shaft at the front of the threads

Fig. 198. Hoisting an engine with a crankshaft lifting eye. (Courtesy Continental Aircraft Engines)

for the thrust-bearing nut. Check for the bottoming of the front cone against the outer ends of the splines by applying a thin coat of Prussian Blue to the inner end face of the front cone, assembling the parts and firmly tightening the nut. Remove the front cone and see whether the Blue has been transferred to the ends of the splines. Clean off the Prussian Blue and reassemble. Bottoming is cause for rejection of the cone.

To install the propeller, assemble the rear cone, propeller or hub, front cone, and propeller-hub retaining nut on the crankshaft in the order named. Screw in the retaining nut and firmly tighten with a bar affording approximately 4 ft. of leverage. One man of average weight (about 175 lb.) using a bar of this length can usually tighten the nut to the proper tension without need of additional leverage or the use of a hammer on the bar. Follow the manufacturer's directions. Install the snap ring and secure the retaining nut by installing and cotter-pinning the clevis pin. The clevis pin should be inserted from the inside of the shaft with a plain washer placed under the cotter pin.

In order to prevent possible bearing failure due to lack of proper lubrication during first starts, the engine should be pre-oiled as follows.

1. Fill the oil tank to the proper level.

2. Break the oil-inlet connection at the oil pump and drain approximately ½ gal. of oil.

3. Reinstall the oil-inlet line to the oil pump.

4. Remove all front spark plugs from the engine.

5. Remove the ⅛-in. pipe plug at the center of the oil pump.

6. Open the throttle to the full-open position.

7. Place the fuel valve in the OFF position.

8. Make sure the ignition switch is off.

9. Turn the engine over by hand until enough oil has come

Fig. 199. The power-plant installation and nose wheel of a light airplane. (Courtesy Engineering and Research Corporation)

through the hole in the oil pump where the plug was removed to show that no air remains in the oil-inlet line or oil pump.

10. Replace the plug in the oil pump.

11. Make several "dummy" starts to obtain a minimum of 30 revolutions of the propeller.

12. Reservice the oil tank.

13. Reinstall the spark plugs.

14. Make a normal start.

The oil pressure may exceed the normal operating pressure when first starting, but it will soon drop to its normal operating pressure as the temperature of the oil increases. As the oil reaches its normal temperature, the pressure should be checked to make sure that it is within the range recommended by the manufacturer. If the oil-pressure gauge does not indicate pressure within 30 sec., the engine should be stopped and not started until the cause of the lack of oil pressure has been determined.

Whenever a new or newly overhauled engine has been installed in an airplane, the engine should be run on the ground for approximately 30 min. During this time, the operation of all engine instruments and the engine itself should be thoroughly checked for proper functioning. Engines running on the ground tend to overheat, and the cylinder-head and oil temperatures must be watched during this run-up period. The engine cowling should be left off during a test of this kind.

Fig. 200. Part of the power-plant installation on a light airplane. (Courtesy Engineering and Research Corporation)

If the engine performs properly during the test period on the ground, it should be flight-tested. The airplane should be flown for not less than 1 hr. The first 50 min. of flight should be at the minimum power necessary for proper operation. The engine should be operated for about 10 min. at its normal rated power. If no defect or improper functioning appears after this inspection, the airplane may be released for service. If the engine does not function properly, the necessary adjustments or corrections should be made, and the airplane again test-flown as before. The mixture control should be kept at the full-rich position at all times during the flight test.

The removal of the propeller, cowling, engine, and engine accessories are just the reverse of the installation procedure.

Engine Run-in After Overhaul. After a major overhaul, the engine should be carefully checked and run-in for a period of from 7 to 10 hr. The length of the run-in depends somewhat on the number of new parts which were installed. The purpose of the run-in is to seat the piston rings and to polish the new parts, such as bearings and pistons. At this time a careful check should be made to ensure that the engine is running satisfactorily and to make any minor adjustments necessary.

An engine run-in should always be under the supervision of an ex-

214

perienced aircraft-engine mechanic. The amount of oil and fuel used, and all other factors connected with the performance of the engine should be carefully checked and noted. The performance of the engine should, if possible, be compared with the performance of a new engine.

Whenever possible, the engine should be run-in on an enclosed test

Fig. 201. A 4-bladed wooden test club installed on an inverted in-line 6-cylinder engine on a test stand. (Courtesy Ranger Aircraft Engines)

stand. If necessary, however, the engine may be run-in while mounted in an airplane. When the engine is run-in on the airplane, it is more difficult to keep a close check on the engine temperature, the fuel and oil consumption, and the engine's overall performance. When a test stand is used, arrangements should be made whereby the carburetor air may be heated, the oil temperature controlled, and the fuel accurately measured. A Cuno type of oil filter and an air cleaner should be installed.

A suitable test club should be used when the engine is run-in on a test stand. A two-blade flight propeller may be used, but is not as desirable as a four-blade club since the regular flight propeller does not furnish as much cooling air as the club. If a two-blade propeller is used, a careful watch on the engine temperature should be maintained.

If the engine is to be run-in on a standard test stand, the following equipment should be installed:

1. A manometer to measure the manifold pressure,

2. A tachometer,

3. A throttle (the mixture control should be wired in the full-rich position),

4. Connections for the oil-inlet and oil-outlet lines,

5. Oil-inlet and oil-outlet thermometers,

6. An oil-pressure gauge,

7. A fuel line to the carburetor, or to the fuel pump if it is mounted,

8. A fuel-pressure gauge in the line to the carburetor,

9. The magneto ground wires connected and high- and low-tension leads from the coils provided for battery ignition,

10. Not less than two, and preferably four, rear spark-plug-gasket thermocouples with their indicating instruments,

11. Ducts to carry temperature-controlled air to the carburetor,

12. A carburetor-air thermometer,

13. For in-line engines, extra air scoops to cool the rear cylinders.

Before starting, whether mounted on a test stand or in an airplane, a newly overhauled engine should be turned over a few times by hand in the direction of normal rotation. This will make sure that there is no excess of oil in the lower cylinders. The manufacturer's directions which are usually given in the operation and maintenance handbook should be followed for engine starting.

If the engine is on a test stand it should be warmed up for at least 15 min. at approximately 800 to 900 r.p.m. Before starting the run-in proper, the oil temperature should be at not less than the minimum operating temperature. The oil-pressure-relief valve should be adjusted to maintain the required oil pressure, and the fuel pressure should be adjusted to that required by that particular engine. When a test club is used, most engines may be run-in at higher r.p.m. than when using a two-blade flight propeller. When running-in on the test stand, one manufacturer recommends 1 hr. at 1200 r.p.m., 1 hr. at 1400 r.p.m., 2 hr. at 1600 r.p.m., 2 hr. at 1800 r.p.m., and approximately 1 hr. at 2000 r.p.m.

If the engine is mounted in an airplane for run-in, all cowling should be left off and the airplane headed into the wind, if there is any, to assist in cooling. Under these conditions, the engine should be operated for 1 hr. at 500 to 700 r.p.m., 2½ to 3 hr. at 1000 r.p.m., and then cut back to idling for about 20 min. Next the engine should be run-in for about 20 min. at 1200 to 1300 r.p.m. and cut back for 2 min. on idling, and these periods alternated until a total of 2 hr. of running time has been completed. The engine then should be run-in at from

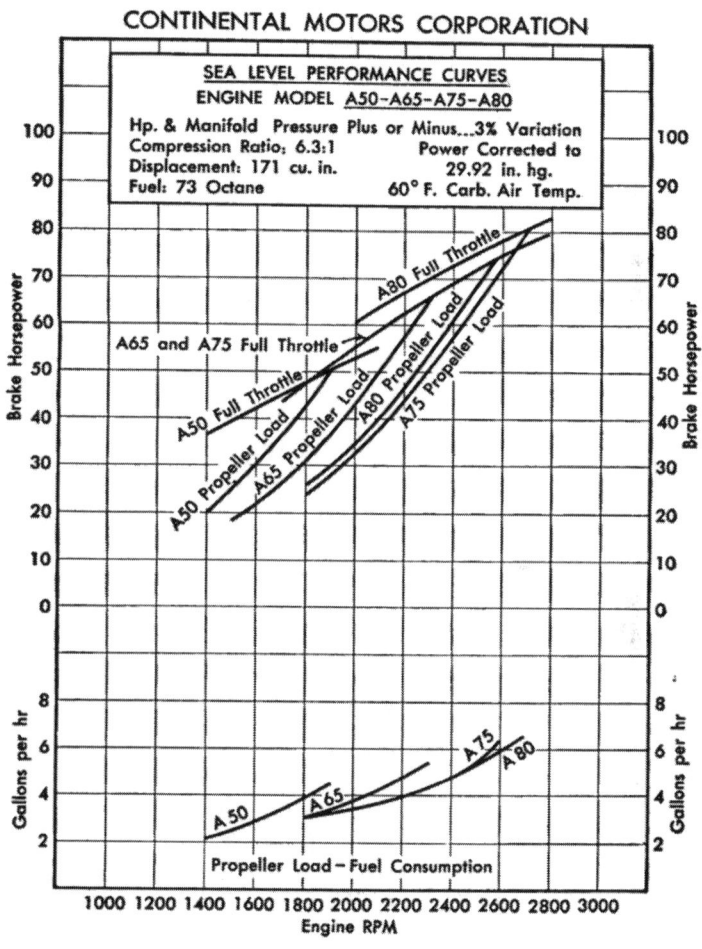

Fig. 202. A performance curve of a light aircraft engine. (Courtesy Continental Aircraft Engines)

1500 to 2000 r.p.m. in short spurts for another 2 hr. During this time, the engine should not be held at 2000 r.p.m. for more than a minute at a time. It should then be cut back to approximately 1500 r.p.m. for 4 min. before running at 2000 r.p.m. again.

If the engine is equipped with a mixture control, the control should be kept in the full-rich position. During the entire run-in, the cylinder temperature, if the engine is equipped with a cylinder-temperature indicator, the oil temperature, oil pressure, fuel pressure, generator charging rate, manifold pressure, carburetor-air temperature, and r.p.m. should be carefully observed and recorded at regular intervals. The barometer reading and outside air temperature should also be noted. The pressure and temperature should always be maintained within the recommended operating limits given by the manufacturer.

When running-in the engine in an airplane, it is of the greatest importance to watch cylinder head and oil temperature closely to prevent overheating.

Whenever possible, it is recommended that, after the run-in with the engine thoroughly protected from dust and dirt, several cylinders be removed for inspection of the cylinder barrel, piston, and piston rings. The manufacturer of one radial engine recommends the removal of cylinders No. 2, 6, and 7. The oil should not be washed from the parts when making this inspection, and the piston rings should be disturbed as little as possible. The piston should not be removed from the engine. If all the parts are in good condition, additional oil should be applied to the cylinder and piston and the cylinder reinstalled. If there is any indication of scoring of piston rings or cylinders, all cylinders should be removed and all necessary repairs made. The engine should then be reassembled and run-in an additional time, either on a test stand or in an airplane. The extra run-in should be at least 3 hr. in length. If it has been necessary to replace any parts, the run-in time should be increased accordingly. The log of the engine run-in and test should be made and filed with the engine inspection and overhaul records or the engine log book.

If the engine is to be placed in storage and not immediately put into service and has been tested on leaded fuel, it should be run for at least 15 min. on unleaded gasoline before it is removed from the test stand or airplane.

Storage of Engines. If an engine is to be placed in storage after the run-in or upon removing from an airplane, the outside of the engine should be thoroughly cleaned with a spray of white furnace oil or other suitable solvent to remove all oil, dirt, and grease. Care should be taken to prevent the cleaning solvent from coming into contact with the electrical equipment.

Moisture often condenses within the rocker boxes and is a common cause for rust at these points. It is recommended that the rocker-box covers be removed while the engine is still warm. Valve clearances should be rechecked when the engine is thoroughly cool. Such parts as cylinder barrels, valves, and bearing surfaces are apt to corrode while in storage if the engine remains idle unless they are given the proper corrosion-preventive treatment. Corrosion frequently occurs even when

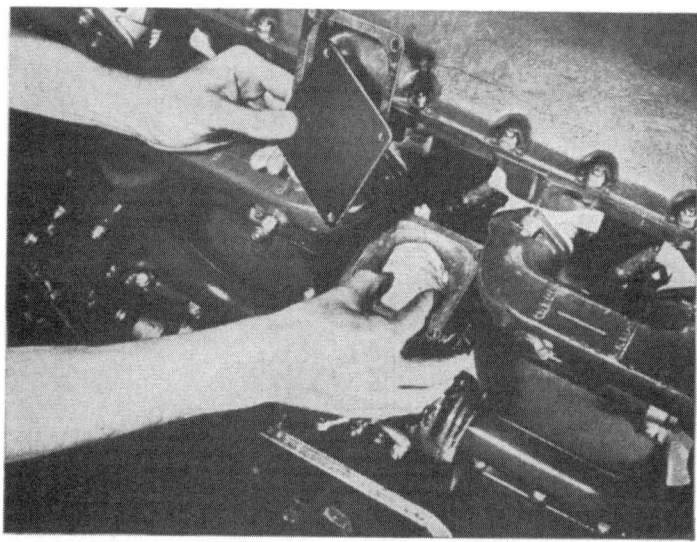

Fig. 203. A bag of silica-gel being placed in the carburetor throat in preparation for storage or shipping. (Courtesy Ranger Aircraft Engines)

Fig. 204. A plastic dehydrator plug being installed in the camshaft-housing front cover. (Courtesy Ranger Aircraft Engines)

engines are stored in a hangar and other enclosed spaces where the conditions are considered favorable. High humidity, rapid changes in temperature, and salt air tend to produce corrosion rapidly.

An engine which is not to be stored for excessive lengths of time should be operated at least once a week to prevent corrosion. The crankshaft should be rotated 6 or 8 times by hand every 3 days whenever an engine is standing idle.

When the engine is to be in storage for a considerable length of time, all parts subject to corrosion should be thoroughly coated with a rust and corrosion preventive recommended by the manufacturer of the engine. The cam and all other parts of the breaker mechanism should be coated with petrolatum. The breaker points should not be coated. The

Fig. 205. A dehydrator installed in the spark-plug opening. (Courtesy Ranger Aircraft Engines)

Fig. 206. A dehydrator plug being installed in the vent hole in the front crankcase section. (Courtesy Ranger Aircraft Engines)

magneto should be wrapped in oiled paper and sealed so that it is dust and moisture proof.

Corrosion preventive should be sprayed into the timing plug opening and into the oil-tank vent connected to the openings, while the crankshaft is being rotated.

The carburetor should be drained of fuel, and flushed with a light oil and completely drained.

In order to remove any vapor which might cause corrosion in the cylinder, the crankshaft should be rotated 6 or 8 times with the spark plugs removed.

The rocker boxes should have the covers removed. On engines having automatic valve lubrication, the operating parts in the rocker boxes should be cleaned by brushing or spraying with white furnace oil and

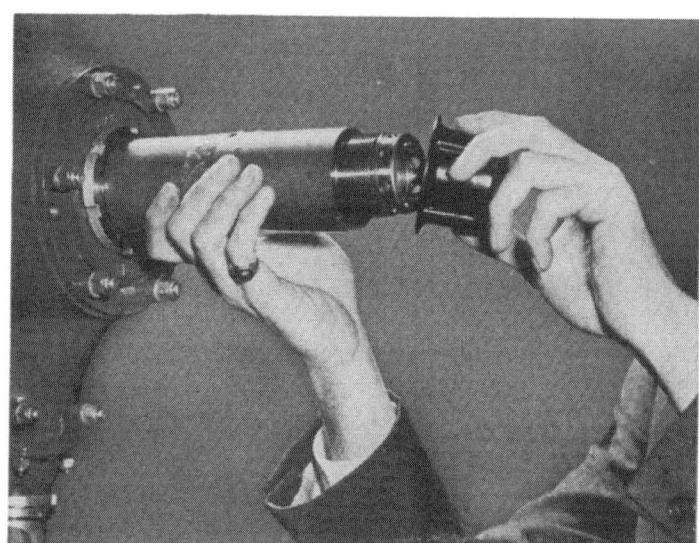

Fig. 207. A protective sleeve and cap placed over the propeller end of the crankshaft. (Courtesy Ranger Aircraft Engines)

then sprayed with corrosion-resistant material. On other engines, the rocker boxes do not ordinarily need to be cleaned of grease, but should be serviced with fresh rocker-arm lubricant.

All accessories should be coated with corrosion-resistant material and, whenever possible, should be wrapped or covered with oiled paper as nearly dust and moisture tight as possible.

With the spark plugs removed, the inside of each cylinder should be sprayed with corrosion-resistant material. The cylinder wall should be sprayed with the piston at bottom dead center on the intake stroke. The crankshaft should be rotated so that the valves can be sprayed when open to make sure that they are thoroughly coated. The crankshaft should be turned two complete revolutions and each cylinder again sprayed with the corrosion-resistant material. After the last spraying, the crankshaft should not be rotated.

The spark-plug hole should be plugged with solid shipping plugs.

Fig. 208. Bags of silica-gel are attached to the sides of the cylinders. (Courtesy Ranger Aircraft Engines)

Fig. 209. The engine is lowered into a pliofilm bag. (Courtesy Ranger Aircraft Engines)

The spark plugs should be given a thin coat of corrosion resistant and wrapped in oiled paper.

All unprotected parts of the exterior of the engine should be sprayed with the corrosion resistant. All oil, grease, and corrosion-resistant compound should be removed from all exposed rubber parts.

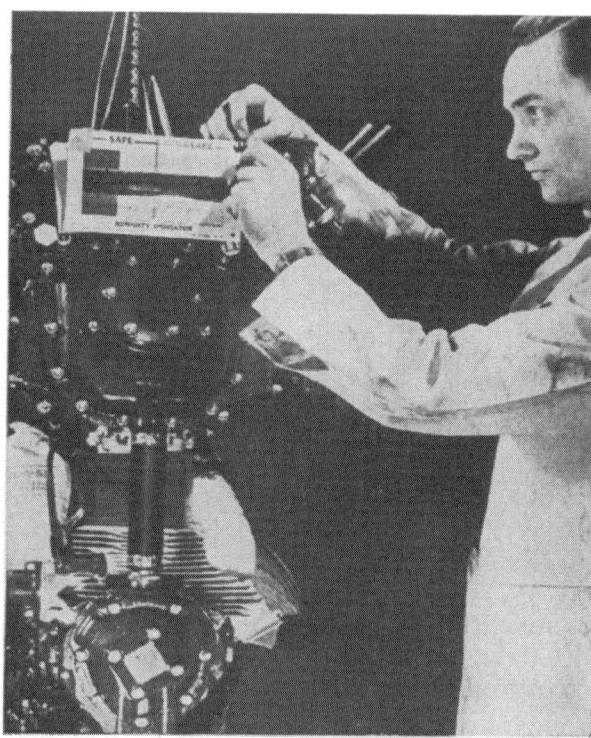

Fig. 210. A humidity indicator is attached to the engine. (Courtesy Ranger Aircraft Engines)

When the engine is to be placed in service after the treatment described above, an effort should be made to remove as much of the corrosion-resistant compound from the crankcase as possible. Most anti-corrosion compounds are soluble in petroleum oils and may be readily removed. Before starting the engine, the oil tank should be approximately half filled with clean engine oil.

Excessive amounts of corrosion-resistant material in the cylinder, intake pipes, or valve ports can cause a bent or broken connecting rod unless they are removed before the engine is started. It is recommended that the intake pipes be removed and the crankshaft rotated 4 or 5 times to make sure that all oil or corrosion-preventive compound drains from the cylinders. Immediately before starting the engine, the propeller should again be turned over at least 3 revolutions by hand to ensure that all excess liquid has been drained out.

The carburetor should be flushed out with gasoline and thoroughly cleaned.

Magneto and distributor parts should be thoroughly cleaned and

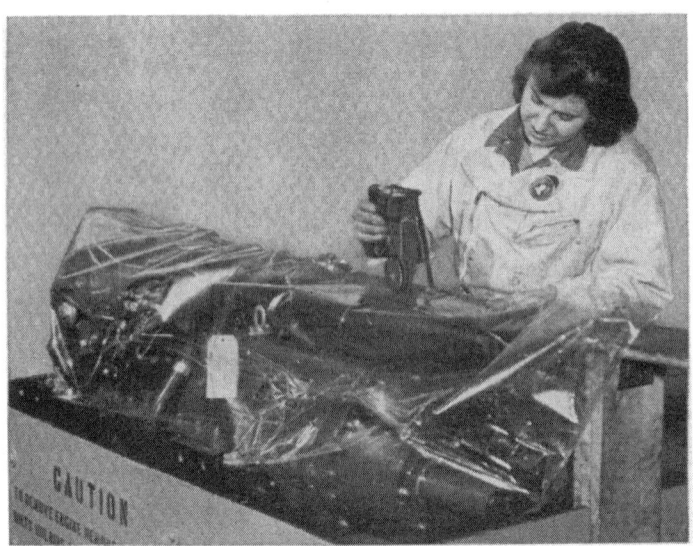

Fig. 211. The edges of a pliofilm bag are sealed with a special heating iron. (Courtesy Ranger Aircraft Engines)

lubricated as required. The spark plugs should be cleaned and reinserted, and the engine operated at not more than 1500 r.p.m. for 20 min. The lubricating oil should then be drained to remove as much of the remaining corrosion-resistant compound as possible. The drained oil should be discarded and not used again. When refilled with clean lubricating oil, the engine is ready for regular operation.

XV PROPELLER MAINTENANCE
AND SERVICE

Propellers are considered to be a part of the power plant. The propeller is one of the most important parts of the aircraft, for upon it depends all of the thrust which pulls the airplane through the air. The Civil Aeronautics Administration has set up very precise and complete rules and regulations pertaining to the certification of propellers and their maintenance and service.

Wooden Propellers. At frequent intervals, before flight and at least once a day, when in operation, wooden propellers should be inspected for any signs of failure or damage. Such defects as cracks, bruises, warping, oversize holes in the hub, scars, evidence of glue failure, or separated laminations or defects in the finish should be repaired at once. If the damage is serious, the propeller should be replaced by another propeller in good condition. The tip or leading-edge covering should be inspected for such defects as separation of soldered joints, loose screws, looseness or slipping, loose rivets, corroded sections, breaks, or cracks.

Causes for Rejection. Any one of the following is sufficient cause for rejecting a propeller for further service.

1. Oversize bolt or hub holes or elongated bolt holes. It is never permissible to plug and rebore bolt holes in a wooden propeller.

2. Excessive number of rivet or screw holes.

3. Any separation of laminations.

4. Any noticeable warping of the blade.

5. Any appreciable amount of the propeller material missing.

6. Wide or deep crack or cuts parallel to the grain of the wood.

7. Any deep cut across the grain of the wood.

8. Any damage to the metal shank of adjustable-pitch wooden blades, such as a crack or cut.

Small Cracks and Cuts. Small cracks and cuts parallel to the grain of the wood should be filled with glue which is worked thoroughly into all portions of the damaged area. After thoroughly drying, the glue should be sanded smooth and flush with the surface of the propeller, and that area of the propeller refinished.

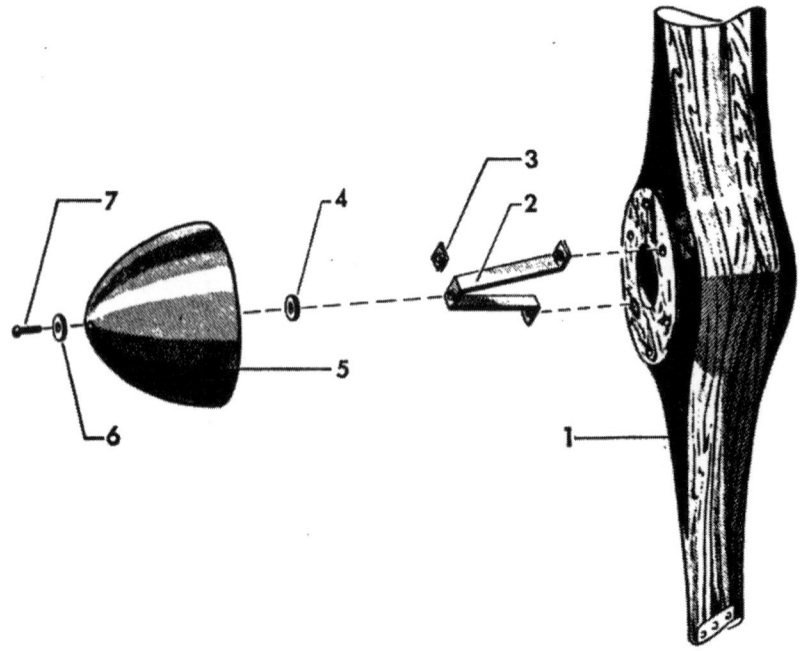

Fig. 212. A wooden propeller installation: (1) Propeller; (2) spinner strap; (3) nut; (4) spinner spacer; (5) spinner cap; (6) washer; (7) screw. (Courtesy Taylorcraft Aviation Corporation)

Dents or Scars. Dents or scars which have rough surfaces or shapes which are slightly rough but not enough to cause failure of the blade may be filled with a mixture of casein glue and fine, clean sawdust. This mixture should be worked and packed thoroughly into the defect. After drying, the area should be sanded smooth and flush with the surface of the blade, and then refinished. All loose splinters should be removed before making the repair.

Damage Requiring Inlays. When portions of the blade are missing and the loss of such portions does not materially weaken the main blade structure, inlays are sometimes allowable. The inlay should be of the same wood as the propeller blade. The material used should be, as nearly as possible, of the same specific gravity as the rest of the blade. The joints should have a taper of not less than 10 to 1 from the deepest point to the feather edge of the inlay. Measurements are taken along a straight line parallel to the grain or general shape of the blade. All hidden repairs should have not less than two uncut coats of high-grade exterior spar varnish before being covered by the metal tipping or

Fig. 213. An adjustable-pitch propeller having aluminum alloy blades built in 1923. (Courtesy Hamilton Standard Propellers)

other material. Repairs that are not under metal tipping should not be covered with fabric or have the wood stained to such an extent that the repair is hidden. The grain of an inlay should extend in the same direction as the propeller laminations. The dove-tail type of joint should not be used on any inlay. The number of inlays on any one blade should not exceed one large, two medium, or four small widely separated inlays. A trailing-edge inlay and a leading-edge inlay should not overlap more than 25 per cent.

Neck, Shank, and Hub. Serious damage to the neck or hub of a wooden propeller is usually cause for rejection of the propeller. Only the smallest inlay should ever be used when there is any question of affecting the strength of the blade. Fairly large repairs are sometimes allowed in the neck and shank where the cross section is fairly large. These repairs, however, are limited to about 5 per cent of the section in thickness. If the neck and shank of the blade are comparatively small, no repairs should be made. Shank inlays, if the shank is large in proportion to the rest of the blade, may be allowed, but should not exceed $7\frac{1}{2}$ per cent of the thickness of the section. Replacements to the hub front and rear face should not exceed 5 per cent of the specified hub thickness.

Blade Repairs. On blades with normal sections from the midsection to the tip, a cross-grain cut up to 20 per cent of the chord in length and $\frac{1}{8}$ of the section in thickness at the deepest point may be repaired. On thin blades, this depth should not exceed $\frac{1}{20}$ of the blade thickness at any point. No repairs to damage of this type should be made in either manner when the feathering edge of the cross-grain cut is closer than 2 in. from the inner edge of the metal tip. The trailing edges may be repaired if the damage consists of narrow slivers up to ⅛ in. wide broken from the trailing edge at the widest portions of the blade. This type of damage may be repaired by sandpapering a new trailing edge. As little material as possible should be removed, and the

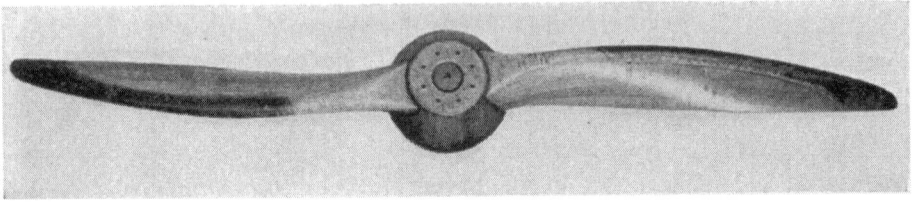

Fig. 214. A wooden propeller used on the Douglas World Cruisers which were flown around the world by the United States Army in 1924. (Courtesy Hamilton Standard Propellers)

repaired portion should blend smoothly into the rest of the blade. Both blades should be balanced by having the same amount of material removed.

A repair near the hub or tip should be made by an inlay. The depth of the inlay edge should not exceed, at its greatest depth, 5 per cent of the chord. The fishmouth joint is the most desirable joint to be used in making an inlay. If the wood of the blade is worn away at the end of the metal tipping, the metal should be removed to make the minimum repair taper of not less than 10 to 1 each way from the deepest point. Due to the convex shape of the leading edge of the average propeller, this taper usually works out about 8:1. Any repair made under metal tipping should not exceed 7½ per cent of the butt or scarf joints and 10 per cent for fishmouth joints. The maximum depth of any repair of this type should not exceed ¾ in.

All finishes applied to wooden blades should be transparent. The finish should be applied in accordance with the manufacturer's recommendations.

Tipping should be replaced when it can no longer be properly repaired. Small cracks in the narrow necks of the metal between pairs of lobes of the tipping commonly occur and are not considered to be defects. All other cracks in the metal tipping are defects and should be repaired, or the tipping replaced. After any repair has been made to a blade, the propeller should be carefully balanced by following the propeller's maintenance as outlined in (CAM–14).

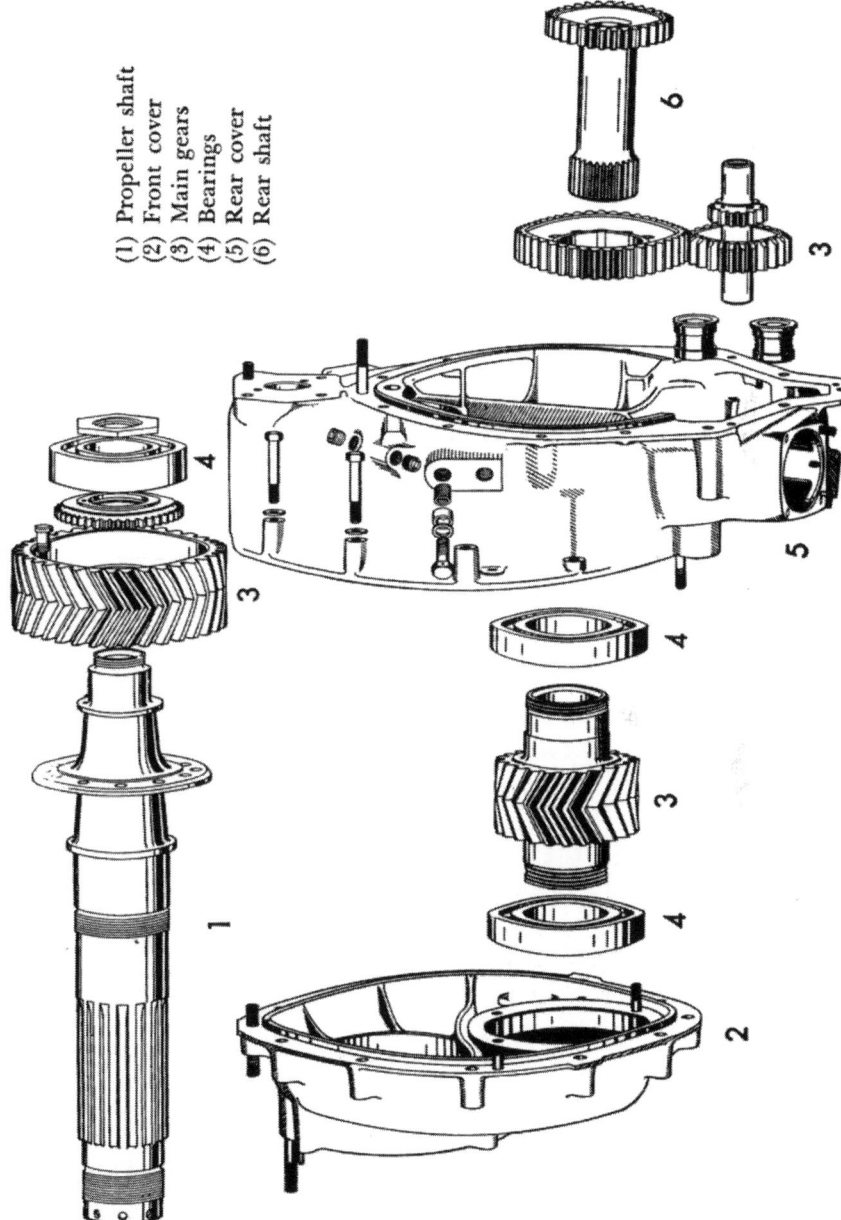

(1) Propeller shaft
(2) Front cover
(3) Main gears
(4) Bearings
(5) Rear cover
(6) Rear shaft

Fig. 215. A propeller reduction gear for an inverted aircraft engine. (Courtesy Ranger Aircraft Engines)

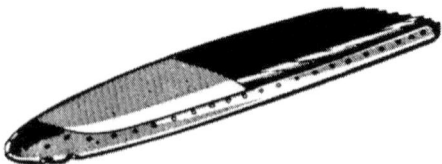

DAMAGED PROP TIP

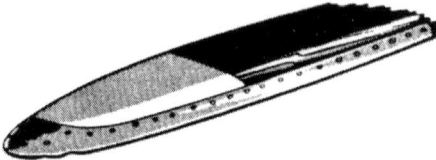

REPAIRED PROP TIP

NOTE:
Round and smooth all
damaged surfaces

Fig. 216. A method of repairing a damaged propeller tip. (Courtesy Taylorcraft Aviation Corporation)

Fig. 217. A phantom view of the reduction gear used on large, radial aircraft engines. (Courtesy Wright Aeronautical Corporation)

Steel Propellers. Steel propellers that have been damaged should not be repaired except by the manufacturer. Minor injuries to the leading and trailing edge, only, of steel blades may be smoothed out by hand stoning, providing the injury is shallow.

Aluminum Alloy Propellers. A seriously damaged aluminum alloy propeller blade should be repaired by the manufacturer or by an approved repair station certificated for this type of work. A metal propeller is considered to have been damaged if it has been bent, cracked, or seriously dented. Minor injuries to the surface, such as small dents, nicks, scars, or scratches which are removable by the maintenance crew, are not considered as damage to the propeller.

Metal Propeller Inspection. At regular intervals metal blades should be given a local etching and examined with a magnifying glass for cracks, or other similar damage. When working with a shallow crack, the worked surface should be etched to be sure that the crack has been entirely removed.

Minor Repairs to Propeller Blades. Any damage in the form of a nick, scratch, scar, cut, or dent which is not serious enough to weaken the blade materially may be removed by the maintenance mechanic. The metal around such damaged areas should be removed, forming

Fig. 218. An exploded view of the propeller reduction gears used on a high-powered, radial aircraft engine. This type of reduction gear has a ratio of about 16:9, the crankshaft rotating 16 times while the propeller rotates 9 times. (Courtesy Wright Aeronautical Corporation)

shallow saucer-shaped depressions which contain no sharp edges or abrupt changes in direction. Such repairs may be made only when the amount of metal removed is not enough to materially weaken the blade. The finished depression caused by the removal of such damage should not be more than 1/8 in. in depth at its deepest point, 3/8 in. in width overall, and 1 in. in length overall. If the depression exceeds any one of these dimensions, the blade is considered to be unserviceable.

Defects Allowable in Blades. Any damaged area which has been repaired constitutes a defect to a propeller blade. A reasonable number of such defects on the blade is not necessarily dangerous. Consideration should be given to the distance between these repairs and their location with respect to each other and to whether or not the total of their effects materially weakens the blade.

Leading Edges. Blades that have the leading edges pitted from normal wear in service may be repaired by removing enough material to eliminate the defect. The metal should be removed by starting at

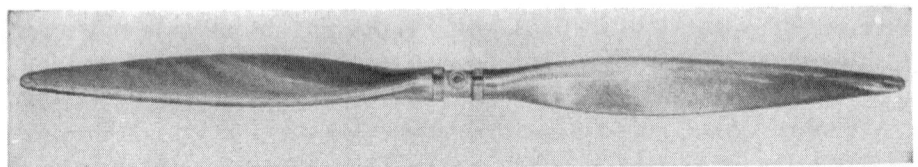

Fig. 219. An adjustable propeller having aluminum alloy blades built in 1924. These blades have threaded ends and are clamped to the hub. (Courtesy Hamilton Standard Propellers)

approximately the thickest section and working forward over the nose camber so that the contour of the reworked portion is approximately the same as the rest of the leading edge. No abrupt change in direction should be made on the leading edge of a blade. The manufacturer's recommendations and the rules of the Civil Aeronautics Administration should be carefully observed.

Shortening of Blades. If it becomes necessary to remove the tip of a blade due to tip damage, the operation should be in accordance with the manufacturer's recommendation and the Civil Aeronautics Administration rules and regulations. If one blade has been shortened, it is always necessary to shorten the other blade in exactly the same manner. Such sets of blades should always be kept together.

Rejection of Propeller Blades. Unless authorized by a representative of the C.A.A. to the contrary, any of the following is cause for rejection of a propeller blade:

1. A transverse crack of any size.

2. The removal of too much of the blade metal by etching or dressing off defects or by shortening the blade.

3. A longitudinal crack, cut, scratch, scar, or other damage that cannot be dressed off or hollowed out without materially weakening or impairing the blade.

4. Any repair which materially changes the cross section of the blade.

5. An excessive number of slag inclusions or cold shuts, or an excessive number of both.

Propeller Governors. The governors or other parts directly concerned with the operation of certificated propellers should be made in accordance with the propeller manufacturer's recommendations. Any replacements to these parts should be made with identical parts produced by the original manufacturer or parts approved by a representative of the Civil Aeronautics Administration. Welding of the propeller governor is not permissible. When a propeller governor has been repaired, it should be tested as recommended by the manufacturer.

Tracking. The tracking of a propeller is an operation which should be performed at least daily before flight. This is the operation by which it is determined whether or not each propeller blade follows the path of the other blade or blades. A bench, stepladder, or other object which may be easily moved is placed just touching one blade of the propeller which is turned to a down position. The propeller is then rotated, and the mechanic determines whether the corresponding point of each blade just touches the same point on the table or bench as the propeller is rotated. Propeller blades do not usually follow exactly the same track, but the maximum allowance for variation between blades, which varies from about $\frac{1}{16}$ in. to $\frac{1}{8}$ in., should not be exceeded. If the blades vary more than the recommended maximum amount, the propeller should be carefully checked to determine the condition of the hub on the shaft or, in the case of a wooden propeller, the propeller hub bolts should be examined. The tracking of a wooden propeller should not be brought into proper adjustment by placing shims between the propeller hub and its mounting. If the propeller is more than the allowable amount out of track, the cause should be determined and corrected or the propeller replaced.

Propeller Hubs. Repairs to propeller hubs and to certificated propellers should be made only in accordance with the propeller manufacturer's recommendations. Welding is never permissible on steel hubs, clevis pins, bolts, or nuts. Any other small parts should be replaced as soon as they show any indication of wear or distortion. Safety-wire or cotter pins should never be used a second time. Hubs and hub parts should be regularly cleaned and inspected. They should be checked by measurements for any signs of distortion. Any hub which is sprung or distorted, or seriously damaged, should be replaced. Steel hubs should be carefully examined for cracks by Magnaflux or other magnetic type of inspection at every major overhaul period. Any crack or indication of cracking is a cause for rejection of a steel hub. Hubs and clamp links should be cadmium plated after they pass inspection. A substitute for cadmium plating consists of a zinc chromate primer covered by a coating of aluminum lacquer.

Splines and cone seats should be carefully inspected for signs of wear. The splines should be checked with a single-key no-go gauge made to plus 0.002 of the base drawing dimensions for the spline land width. If the gauge enters more than 20 per cent of the spline area, the hub should be rejected.

The maintenance and service of controllable pitch propellers should be carried out in accordance with the "Manufacturer's Maintenance and Service Manual" for the type propeller being serviced.

INDEX

INDEX

The Aviation Collection by Sportsman's Vintage Press

www.SportsmansVintagePress.com

Aircraft Construction Handbook	by Thomas A. Dickinson
Aircraft Sheet Metal Work	by C. A. LeMaster
The Aircraft Apprentice	by Leslie MacGregor
Aircraft Woodwork	by Col. R. H. Drake
Aircraft Welding	by Col. R. H. Drake
Aircraft Sheet Metal	by Col. R. H. Drake
Aircraft Engines	by Col. R. H. Drake
Aircraft Electrical and Hydraulic Systems, and Aircraft Instruments	by Col. R. H. Drake
Aircraft Engine Maintenance and Service	by Col. R. H. Drake
Aircraft Maintenance and Service	by Col. R. H. Drake

Printed in Great Britain
by Amazon.co.uk, Ltd.,
Marston Gate.